**Verdi Community Library
and Nature Center**

CALIFORNIA NATURAL HISTORY GUIDES

INTRODUCTION TO HORNED LIZARDS OF NORTH AMERICA

California Natural History Guides

Phyllis M. Faber and Bruce M. Pavlik, General Editors

Introduction to

HORNED LIZARDS of NORTH AMERICA

Wade C. Sherbrooke

UNIVERSITY OF CALIFORNIA PRESS
Berkeley   Los Angeles   London

As with the first edition, this second edition is dedicated to the many people from around the world and over generations who have created the knowledge and understanding, through observation, experimental study, careful thinking, and publication, upon which this book is based.

California Natural History Guides No. 64

University of California Press
Berkeley and Los Angeles, California

University of California Press, Ltd.
London, England

© 2003 by Wade C. Sherbrooke

Library of Congress Cataloging-in-Publication Data

Sherbrooke, Wade, C.
    Introduction to horned lizards of North America / Wade C. Sherbrooke.
        p. cm.—(California natural history guides; no. 64)
    Rev. ed of: Horned lizards, unique reptiles of western North America. c1981.
    Includes bibliographical references (p.).
    ISBN 0-520-22825-1 (hardcover : alk. paper)—ISBN 0-520-22827-8
(paperback : alk. paper)
    1. Horned toads—West (U.S.)   2. Reptiles—West (U.S.)   I. Sherbrooke,
Wade C. Horned lizards, unique reptiles of western North America.   II.
Title.   III. California natural history guides ; 64.

QL666.L25 S48 2003
597.95—dc21

                                                                2002038745

12   11   10   09   08   07   06   05   04   03
10   9   8   7   6   5   4   3   2   1

The paper used in this publication meets the minimum requirements of
ANSI/NISO Z39.48-1992 (R 1997) (*Permanence of Paper*). ♾

The publisher gratefully acknowledges the generous
contributions to this book provided by

the Moore Family Foundation
Richard & Rhoda Goldman Fund
and
the General Endowment Fund of the University of
California Press Associates.

CONTENTS

PREFACE

The first edition of this book was published in 1981 as *Horned Lizards: Unique Reptiles of Western North America* by the Southwest Parks and Monuments Association. Twenty years later, the book has been revised and expanded to incorporate new knowledge and interpretations and to include species not within the geographic scope of the first edition. It now covers all species of lizards in the morphologically distinct genus *Phrynosoma*, horned lizards.

Throughout the text, I attempt to provide insights into the complexities of the lives of these animals without resorting to unexplained technical terminology. The first edition was popular with both adults and older children, and I hope this one will be, too. Discussions are supported by illustrations, usually photographic. It is important for the reader to see the fine details of the lives of these lizards if he or she is to understand what horned lizards are all about, and their origins and ecological existence on earth today. As another life form on earth, we humans have much in common with them.

I hope that this book will find wide use not only among professional and amateur herpetologists—people who study and enjoy reptiles and amphibians—but also among a much wider public of readers who are searching for ways to expand their personal engagement with the natural world and their connection to it. In recent years, horned lizard designs have been incorporated into an expanding array of artistic works, particularly in the geographic regions where these lizards occur. As awareness of the lives of these lizards increases, perhaps people entering their territory, as outlined on the distribution maps, will take note. Perhaps they will see something other than the "wasteland" reported by some earlier travelers, namely, a life-support system for a fellow creature that they have come to understand and value.

ACKNOWLEDGMENTS

While writing about a subject that has been a primary focus of my life for 25 years, it became clear how broad and deep my debts and gratitudes extend. This book could never have been completed without the help of a myriad of people. Each person has stimulated my understanding and growth, and I owe a part of this book to all of them, even if their names do not now come immediately to mind.

Clearly, my debts go back to the first edition, and again I thank all those responsible for their encouragement, inspiration, and assistance with that book. Those early days were critical in forming my interests and confidence. My interest in coloration and color change was fostered during my dissertation studies by Joseph T. Bagnara, Ana Maria de L. Castrucci, and Mac E. Hadley.

Since 1985 I have lived at the Southwestern Research Station, a field research station of the American Museum of Natural History, in the Chiricahua Mountains, southeastern Arizona. I have had the support of the museum for much of my work on horned lizards, as well as the support and interest of scientific researchers, volunteers, students, and naturalist guests who visit the station. They have all shared generously their knowledge and curiosity.

George A. Middendorf III has been a delightful colleague on investigations into the strange defense of squirting blood. Charles R. Bryant, a geologist, brought a dead rattlesnake to me with a horned lizard fatally lodged in its throat. Harry W.

Greene and David L. Hardy Sr. introduced me to the excitement of radiotelemetry, a technique that has greatly enhanced my efforts at understanding these lizards. Barney Tomberlin and Mitch Webster have helped me keep track of the activities of horned lizards in the boot heel of New Mexico. My understanding of the Flat-tail Horned Lizard *(Phrynosoma mcallii)* was enhanced by association with Wendy L. Hodges, Bryan L. Morrill, Kevin V. Young, and April Young, and my understanding of the Desert Horned Lizard *(P. platyrhinos)* was enhanced by Dale Turner. Kelly R. Zamudio and Harry Greene introduced me to the Pygmy Horned Lizard *(P. douglasii),* and Kelly has been an inspirational horned lizard colleague ever since. Other herpetologists to whom I feel particularly indebted are G. Larry Powell, Philip J. Fernandez, and Kurt Schwenk. My explorations in Mexico were greatly facilitated by the kindness and experience of Fernando Mendoza Quijano and Sol de Mayo Araucana Mejenes Lopez and by Elizabeth Beltran Sanchez, David Lazcano Villarreal, Mario Mancilla Moreno, Richard R. Montanucci, and Walter Schmidt Ballardo.

Government agencies who have granted me permission to conduct research on horned lizards include the Arizona Department of Game and Fish, California Department of Fish and Game, New Mexico Game and Fish Department, Texas Parks and Wildlife Department, and U.S. Fish and Wildlife Service.

Members of the Horned Lizard Conservation Society have provided a continuing bond of interest regarding the conservation of these lizards and efforts to reach the public on their behalf. Many members have enhanced my knowledge. I am particularly indebted to Larry Wisdom for the series of replicas of horned lizards he has developed as educational tools.

I greatly appreciate Raymond A. Mendez for allowing me to use his remarkable photograph of a Texas Horned Lizard *(P. cornutum)* squirting blood out of its eye socket taken during a session staged by George Middendorf and myself (page 109).

I took all other photographs. Susan Nagoda Bergquist and Betty Manygoats allowed me to use my photographs of their horned lizard ceramic sculptures (pls. 130, 132). For allowing me to photograph materials in their collections, I thank the Arizona State Museum, Tucson (five Hohokam pieces), the Amerind Foundation, Dragoon, Arizona (two Casas Grandes ceramic pots), Mesa Verde National Park (an Anasazi ceramic shard), the Laboratory of Anthropology, Albuquerque (a Mimbres ceramic bowl), and Mark Bahti (Tohono O'odam fetish).

Critical to the realization of this revised edition was an effort made by Harry Greene to introduce me to the University of California Press and to their Executive Editor, Doris Kretschmer. Their encouragement was the catalyst that set me to work. My wife, Emily E. W. Sherbrooke, has lent her valuable editorial skills toward honing my efforts. I thank her for that and her many other supporting activities during the revision process. And I thank our children, Skylar and Reed, for their constant interest in the natural world and all the questions they ask that help keep me in touch with it.

So-called horny toads are actually horned lizards. The confusion evident in these conflicting common names has arisen because these lizards bear a superficial resemblance to toads, a fact clearly recognized by the zoologist who gave the scientific name *Phrynosoma* to the genus of horned lizards in 1828. He named them the "toad-bodied" lizards; in Greek, *phrynos* means toad, and *soma* means body.

Like many toads, horned lizards have a broad body, rough skin, and a rather awkward gait (although they do not hop). Also, in toadlike fashion, they flick out their tongues to pick up insect food and, when molested, inflate themselves with air. But in spite of these toadlike qualities, they are lizards. Like other lizards, they have tails, scales covering the body (rather than moist glandular skin), forefeet with five-clawed toes (as opposed to the four unclawed foretoes of toads), eggs specialized for development on land (they would suffocate under water), and many other traits that confirm their identity as reptiles rather than amphibians.

On first encounter, the bizarre form of horned lizards can conjure up images of long-extinct reptilian giants. In fact, in the 1960 Hollywood film fantasy *The Lost World*, horned lizards "terrified" the actors (and perhaps the audience) when magnified to portray prehistoric horned dinosaurs. But people who are familiar with these lizards know their docile, inoffensive nature.

The distribution of these distinctive lizards in America's deserts has caused them to become symbolic of regions in western North America that have little rain, clear skies, and extreme heat. But the diversity of *Phrynosoma* species that has evolved on this one continent, and no other, can be found in low- and high-elevation habitats in an area that ranges from southwestern Canada through the western United States and much of Mexico to the Guatemalan border.

Although horned lizards are easily recognized as strange and otherworldly, few people appreciate the close ecological relationships between these lizards and our North American landscapes or the means by which they survive in these often harsh environments: sleeping beneath the sand or up in shrubs at night; changing color while basking in the morning sun; hunting and eating venomous ants; seeking and signaling appropriate mates; defending themselves against coyotes and foxes, roadrunners and shrikes, venomous and nonvenomous snakes, and various species of hawks; seasonally hibernating underground; and harboring intestinal parasites. Many of the horned lizards' activities and adaptations for survival are at least as interesting and awe-inspiring as their appearance.

Over the centuries, these creatures have influenced the mythology and art of prehistoric and historic peoples of North America. Their place in nature has always been, as it still is for us today, worthy of contemplation.

TIME AND A CONTINENT

Through their genetic heritage, horned lizards, like humans, are linked to the first Paleozoic land vertebrates and their evolutionary advances. Much more recently, the evolution of their closest reptilian predecessors, and of horned lizards themselves, has been molded by Cenozoic geologic and climatic events throughout North America. Today, 13 currently recognized species of horned lizards live within, and are adapted to, the contemporary diversity of ecological settings on this continent. The lives of these lizards, like ours, are edges of the living past cutting into the ever-arriving future.

Origins

Five hundred million years ago, all the backboned ancestors of horned lizards lived in the sea. About 350 million years ago, more recent ancestors evolved the ability to climb from the water's edge onto land. Soon their descendants, primitive swamp-living amphibians, were supporting themselves on four legs. These limbs, having five terminal digits, were held rigid by the same bones that we find in the limbs and digits of horned lizards and ourselves today.

Those early, air-breathing amphibians had only begun the developments that vertebrates would continue to refine as they became increasingly adapted to life on land. Along with other advances, early reptiles—or their now-extinct amphibian predecessors—solved two major problems limiting amphibian life on land. They evolved skin that reduces the loss of water to the air; and they evolved a new kind of water-containing shelled egg that protects the growing embryo and allows it to develop in a terrestrial environment.

With these advances, reptiles became the dominant vertebrates on land, at sea, and in the air during the age of dinosaurs, the Mesozoic, roughly 250 to 65 million years ago. Some of those giants carried horns and spines, like horned lizards today. An example is *Triceratops,* a formerly abundant, three-horned dinosaur, which was 20 feet long and twice as heavy as a rhinoceros. But the time of the dinosaurs has passed, and change has brought the evolution of birds and mammals. Nevertheless, even today, in the age of mammals, there are nearly 7,000 living species of successful modern reptiles, of which 3,800 are lizards.

Horned lizards are members of the Iguanidae, one of the largest of the 17 families of living lizards. The types of iguanids are varied and include over 50 genera with more than 500 species, almost all living in North and South America. Some herpetologists divide the iguanid family into eight different families (making 24 families of lizards), with horned lizards in the Phrynosomatidae, where they are grouped with sand lizards, tree lizards, fence lizards, zebra tails, and others.

Places

Geologists have determined that about 200 million years ago, the earth's continents were united as one land mass, Pangea, which later broke apart into northern and southern land masses, Laurasia and Gondwana, respectively, each of which later split and slowly moved over the earth's surface, forming the continents as we know them today. These continents are still moving, riding on geologic plates. When these geologic plates crash into each other, they push up mountains onto the seafloor and into the air. These mountains change the flow of ocean currents and wind patterns, thus altering climatic patterns and the nature of life on land. In western North America, this process is relatively recent and ongoing, and the signs of geologic instability are obvious—earthquakes, rugged and unweathered mountain ranges, recent lava beds, and active volcanoes. In the western United States and throughout northern, central, and southwestern Mexico, many regions are characterized by aridity. Within these dry plains, basins, plateaus, deserts, mountains, and seasonally dry forests that stretch southward between about 50 and 15 degrees north latitude, across the Tropic of Cancer, horned lizards have found diverse ecological homes.

Exactly when higher temperatures and reduced precipitation formed the most recent dry climates in North America is uncertain, but undoubtedly, the change did not happen all at once. Apparently, it has been going on for nearly 20 million years—more or less the time, gauged by fossils, that horned lizards are thought to have existed as distinct animals.

Today, the part of North America inhabited by horned lizards lies within the more northerly of two broad belts of subtropical deserts found around the northern and southern hemispheres in the vicinity of the Tropics. Within these belts, cold, dry air from high altitudes tends to descend toward the earth's surface and increases in temperature as a result of compressional heating. On arrival at the surface, the warmed air contains little water vapor. Solar radiation easily penetrates this clear air and heats the terrain below to high temperatures. As the earth changes its tilt toward the sun during its annual orbit of our star, climatic patterns rhymically shift back and forth, north and south, creating seasons.

Along most of the western coast of North America, westerly trade winds from the Pacific Ocean are forced up and over the Cascade Range, Coast Ranges, Sierra Nevada, and Sierra Madre Occidental. As the air rises, it cools, and precipitation occurs in the mountains. After losing its moisture, the air descends to the east, where its dryness leaves dry and arid conditions, called "rain shadows." The seasonal timing of scant precipitation in arid areas is important to horned lizards because it can determine the seasonal abundance and relative availability of food and the suitability of nest sites. These rainfall patterns vary regionally.

Equally variable are temperature conditions, which are important to reptiles such as horned lizards. Some regions are colder because they are farther north, at higher elevations, or isolated from the influence of warm oceanic waters near the coast. Dramatic temperature changes are associated with mountainous elevational changes. At higher elevations, temperatures are lower and precipitation is greater. These environmental differences limit the distributions of species of horned lizards and other organisms. Throughout North America, plant species are so consistently limited by climate that ecologists regularly use past and present plant communities as indicators of regional climate.

During the last 1.5 million years, northern North America has experienced four episodes of extensive glaciation known as "ice ages." Because of climatic cooling, massive sheets of ice extended southward from the pole and briefly covered portions of Canada and northern regions of the United States, and ice formed at high elevations in Mexico. Between each glaciation, the climate warmed and the ice receded. Fossils of horned lizards that appear to be direct ancestors of living species of horned lizards have been found in materials surviving from those climatically unstable times. Other horned lizard species in the fossil record have no modern counterparts.

During periods of climatic change, the elevational limits on plants and animals are altered. A species might become extinct in areas where it formerly thrived, or it might spread to places where it could not have survived before. For example, the present elevational distribution of plant species and communities on mountainsides in the southwestern United States is probably 8,000 to 11,000 years old. Before then, and throughout the ice ages, major changes in climate caused communities of plants and animals to shift up and down mountain slopes, as well as northward and southward. The ancestors of today's horned lizards were subjected to the same climatic changes, shifting each species' distribution north and south or up and down mountains. Some populations of a species of horned lizard became isolated from one another by mountain ranges, unsuitable habitats, or simply distance.

As horned lizard populations adapted to local conditions of climate, food, and predators, they became distinct from one another. If genetic differences became significant, they evolved into related populations, subspecies, or even separate species. In contrast, less adaptable species could not make the adjustments demanded by the changing times and became extinct. As a result of these evolutionary forces, the species of horned lizards living today are well adapted to the diverse climatic and ecological conditions found throughout the western United States and much of Mexico. In the north, horned lizards may experience harsh, freezing winters; further south, they may live on blisteringly hot sand dunes, in seasonally dry tropical forests, or within view of active volcanic peaks. Each species is unique, yet all are clearly recognizable as that distinct animal many call the horny toad or horned lizard—or in Mexico, *camaleón.*

DIVERSITY OF A FORM

All horned lizard species are immediately distinguishable from other North American lizards by their horns and flattened bodies. Their ancestors evolved this body form and appropriate behaviors, a mode of life distinct among lizards. Horned lizards have diversified into a number of closely related species of similar form, an evolutionary process known as adaptive radiation. The distinctive life and unique design of horned lizards have been a successful evolutionary development that gives us a window into understanding the processes of life on earth, of which we are a part.

Form

The unusual appearance of all horned lizards is due to their unique and integrated set of adaptations. Their camouflaging colors, flat body form, and sedentary nature make them difficult to see. Their horns and spiny scales make them difficult to swallow. Their pancake shape (pl. 1) provides them with a large, flat surface that is useful for rapid solar heating, capturing raindrops, or warding off predators. Their capacious body provides room for development of numerous eggs or young, depending on the species, and for a large stomach. At the same time, their awkward shape limits their speed, which affects their ability to hunt prey and escape predators. Their shape also makes walking through dense stands of ground-level herbs and grasses difficult. These lizards look strange because some of their methods for survival are contrary to those that we typically associate with lizards. But for horned lizards, these unorthodox strategies for survival have worked, with the benefits outweighing the costs. They continue to survive and reproduce successfully.

Plate 1. Alizarin red staining of skeletal calcium in the Regal Horned Lizard *(Phyrnosoma solare)*.

Diversity

Adaptive radiation, or the differentiation of a basic body plan and survival strategy, has led to the evolution of 13 currently recognized living species of horned lizards. They inhabit areas as far south as the Mexican-Guatemalan border and as far north as southwestern Canada (two species). All but one of the species occur in Mexico, and eight species occur in the United States. Over the years, scientists have reconsidered the number of existing species and have designated some types of horned lizards as formal subspecies, or geographic races. These subspecies are indicators of evolutionary divergence. The subspecies of any one species are still loosely related so that individuals from different subspecies are thought to be able to mate and produce fertile young, as do the various races of humans. But subspecies have differentiated from one another morphologically by adapting to regional ecological conditions. Given time and the proper evolutionary forces, subspecies may evolve into distinct species no longer capable of interbreeding. From that point in time forward, the evolutionary lines of the species are separate from one another. Recent studies utilizing mitochondrial DNA suggest that the genetic differentiation among populations over time may not always correspond to the manifest physical features among formally recognized subspecies of horned lizards. Thus, the significance of differences among populations in some species and their designation as species or subspecies remain in question and are topics for future study.

One interpretation of the evolutionary history of the genus of horned lizards is that the ancestral group separated into what later became a more or less southern group of species, or a southern radiation, and a more northern-distributed group of species, or a northern radiation. The progenitors of these two groups may have been similar to some of the descendant species alive today: the southern Giant Horned Lizard *(Phrynosoma asio),* the Mexican-plateau Horned Lizard *(P. orbiculare),* and the northern Coast Horned Lizard *(P. coronatum).* We can trace these branching radiations by employing a taxonomic tool called a cladogram (see the diagram of hypothesized relationships), which arranges pairs or groups of species who share common ancestors in a branching phylogenetic (family) tree, with time moving from left to right.

Thus, in the northern radiation (all egg-laying species), five xeric-adapted, or desert-adapted, species evolved from an ancestor of the Coast Horned Lizard, which itself may have derived from a species similar and ancestral to the Mexican-plateau Horned Lizard. First, an isolated eastern population split off, perhaps because of the continental uplifting of the Sierra Madre and the Rocky Mountains, and became today's Texas Horned Lizard *(P. cornutum)*. Then the ancestral population gave rise to the Regal Horned Lizard *(P. solare)* (in portions of the developing Sonoran Desert), the Flat-tail Horned Lizard *(P. mcallii)* (in the region of the Colorado River delta), and the Desert Horned Lizard *(P. platyrhinos)* (which spread northward into the Great Basin Desert). The population ancestral to the Desert Horned Lizard may also have spread eastward to become the Roundtail Horned Lizard *(P. modestum)* of the Chihuahuan Desert.

The southern radiation (having all but one live-bearing species), perhaps with its origin in ancestors of the Giant Horned Lizard, the Mexican-plateau Horned Lizard, or both, consists of two species with distributions limited to southern Mexico (Bull and Short-tail Horned Lizards [*P. taurus* and *P. braconnieri,* respectively]) and three species with short horns (Rock, Short-horned, and Pygmy Horned Lizards [*P. ditmarsi, P. hernandesi, and P. douglasii,* respectively]) that are found in northern Mexico or northward, often at higher elevations.

Our understanding of the relationships among living species is still being debated and therefore is speculative. Certainly there were other species of which we have no record. And the data we have from living species, even with mitochondrial DNA studies, do not yield one clear interpretation of past evolutionary events. Horned lizard fossils are few and not particularly helpful in resolving relationships among living species. So, our present-day hypothesis of past evolutionary events may well be modified with future studies; the branching pattern of the cladogram may change, or species may be added to it.

Identity

Geography is a very useful tool when you are trying to identify a species of horned lizard (see the map of distribution). Any location has a limited number of species; no region has more than four. Within these geographical limits, habitat preferences of species can be helpful in locating and identifying individuals. No species occurs in all habitats within its range. Range maps show the historical distribution of a species, including some areas from which it may have been extirpated as the natural world has succumbed to human activities in recent times.

Although all horned lizards resemble one another, a few specific external characteristics enable us to distinguish among species (see the flow diagram for identification). Other features may look different at first examination, but because of variation among individuals within a species, they may not be as useful in distinguishing species from one another. Developing familiarity with some species, by observing them in the field or viewing photographs, will aid in forming a visual mental impression of the species to which details can be added for positive identification.

Particularly useful for recognizing the specific species of horned lizard are differences in the number, length, arrangement, and shape of the horns along the back of the head. Horns near the centerline are referred to as occipital horns, and those to the sides of the head are termed temporal horns. A silhouette of the head and horns typical of each species is used throughout this book (see flow diagram for identification and individual head silhouettes in the species accounts). Each silhouette is a two-dimensional simplification of three dimensions. It is representational for each species, so individuals can be expected to vary from the species silhouette. Other head spines, smaller than horns and not depicted in the silhouettes, sometimes occur above the eyes and along the lower jaw and also vary among species. For some species, the silhouettes are very distinctive in form, whereas for other species, the silhouettes of two species closely resemble each other. In these latter cases, as in all identifications, other identifying characteristics and data such as locality of origin (see map of distribution and species accounts) need to be incorporated into the identification process.

Along the side of the body, most species of horned lizard have a lateral fringe of enlarged spiny scales. Depending on the species, this fringe may be arranged in one or two rows, or it may be absent (see flow diagram for identification). Species also differ in the abundance and prominence of elongated spiny scales on the upper surface of the body. Belly, or ventral, scales are much flatter than those on the back. Each ventral scale may be smooth or keeled (with a central ridge, like the keel under a sailboat), depending on the species.

Certain aspects of color pattern offer consistent and therefore reliable aids for identification. The presence or absence of a line down the middle of the back, and its color, is a distinguishing characteristic among species. Species also differ in their pattern: the size, arrangement, shape, and color of spots or wavy bands on the back. Belly coloration may also be distinct and useful. But certain cryptic colors, particularly blacks, tans, reds, and yellows, can vary strikingly among populations of one species, making them look different. These colors help the lizards blend with the soils or rocks of their particular location.

In four species, the presence or absence of black dots along the front edge, or lip, of the anal (cloacal) opening, or vent, is species specific and an aid to identification (pls. 2, 3). The vent is a transverse slit on the midline under the rear of the body, just behind the attachment of the hind legs (pls. 2, 3). It serves for the elimination of digestive tract and kidney wastes and for copulation and internal fertilization. Because many species of lizards easily lose a terminal portion of their tail, the size or body length of a lizard is considered to be a measurement from the tip of the snout to the vent. Horned lizards,

Plate 2. Enlarged scales behind the vent (anal opening) of the male, with no black vent spotting (Flat-tail Horned Lizard [*Phrynosoma mcallii*]).

Plate 3. Lack of enlarged scales behind the vent (anal opening) of the female, with black vent spotting (Desert Horned Lizard [*Phrynosoma platyrhinos*]).

however, do not easily lose their tail when grabbed by it, nor do they regenerate a new one if the tip is lost. Three species found in Mexico carry a naturally abbreviated tail, which aids in their identification (pl. 4).

In addition to determining the species of an individual horned lizard, you can also externally ascertain its sex. Typically, the male has a few enlarged, flat scales on the lower side of the tail just behind the vent (pl. 2), whereas the female lacks these enlarged scales (pl. 3). The male also has distinctive

Plate 4. A very short tail (Bull Horned Lizard [*Phrynosoma taurus*]).

Plate 5. Femoral pores on the underside of the male's hind leg (Flat-tail Horned Lizard [*Phrynosoma mcallii*]).

structures, known as femoral pores, along the lower back edge of the thigh (pl. 5). In the female, these are smaller and barely visible. The function of these pores in the male horned lizard is still a mystery. One idea is that they leave a scent on rocks for other lizards to detect with their tongues. Another idea is that the male horned lizard's pores secrete a compound that reflects ultraviolet light, which we cannot see but that lizards may see, and that it is left on surfaces as a visual signal. The third external difference between the male and female horned lizard is the shape of the tail base, just behind the vent. In the male this region is much broader than in the female (pls. 2, 3) because it holds the two hemipenes, the copulatory organs (pl. 6).

Plate 6. Two male copulatory organs, hemipenes, extended (artificially) from the vent (Regal Horned Lizard *[Phrynosoma solare]*).

IDENTIFICATION AIDS

Hypothesized Relationships of Living Species

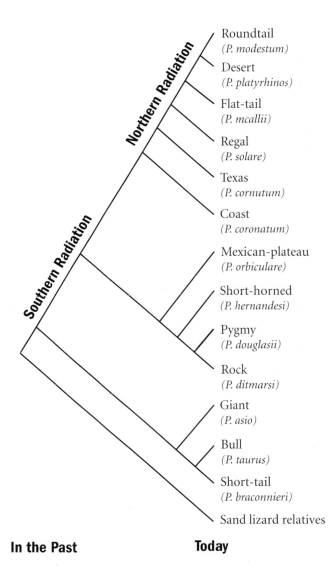

Northern Radiation

Roundtail
(*P. modestum*)

Desert
(*P. platyrhinos*)

Flat-tail
(*P. mcallii*)

Regal
(*P. solare*)

Texas
(*P. cornutum*)

Coast
(*P. coronatum*)

Southern Radiation

Mexican-plateau
(*P. orbiculare*)

Short-horned
(*P. hernandesi*)

Pygmy
(*P. douglasii*)

Rock
(*P. ditmarsi*)

Giant
(*P. asio*)

Bull
(*P. taurus*)

Short-tail
(*P. braconnieri*)

Sand lizard relatives

In the Past   **Today**

Flow Diagram for Identification

Start at left and follow lines/arrows to right while tracking diverging character choices.

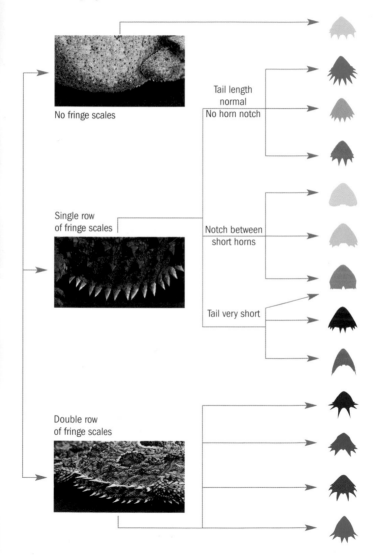

No fringe scales

Tail length normal
No horn notch

Single row of fringe scales

Notch between short horns

Tail very short

Double row of fringe scales

Identifying Characteristics

Always check Map of Distribution to see possible species occurring in area

- ◼ Adults small; four horns of equal size/spacing; black spots on vent

- ◼ Bases of flat horns in contact with each other; no black vent spots

- ◼ Tail not short; ventral scales smooth

- ◼ Horns well developed; black spots on vent

- ◼ Adult size small; horns very short; ventral scales smooth

- ◼ Horns short; ventral scales smooth

- ◼ Back of jaw deep; horns very short; ventral scales keeled

- ◼ Ventral scales keeled; temporal horns separate from occipitals

- ◼ Temporal horns fused into bull-like horns; ventral scales keeled

- ◼ Dark line down back; horns long/sharp; vent without black spots

- ◼ White line down back; dark lines on face and head

- ◼ Dark bands crossing back; bases of occipital horns not in contact

- ◼ Strong eye spines; middle horns often erect; ventral scales keeled

Horned Lizard Species

*Found only in Mexico
See photographs, pp. 22–23

- ◼ Roundtail (pl. 17)
  Phrynosoma modestum

- ◼ Regal (pl. 21)
  Phrynosoma solare

- ◼ Mexican-plateau* (pl. 29)
  Phrynosoma orbiculare

- ◼ Desert (pl. 19)
  Phrynosoma platyrhinos

- ◼ Pygmy (pl. 11)
  Phrynosoma douglasii

- ◼ Short-horned (pl. 13)
  Phrynosoma hernandesi

- ◼ Rock* (pl. 27)
  Phrynosoma ditmarsi

- ◼ Short-tail* (pl. 25)
  Phrynosoma braconnieri

- ◼ Bull* (pl. 31)
  Phrynosoma taurus

- ◼ Flat-tail (pl. 15)
  Phrynosoma mcallii

- ◼ Texas (pl. 7)
  Phrynosoma cornutum

- ◼ Coast (pl. 9)
  Phrynosoma coronatum

- ◼ Giant* (pl. 23)
  Phrynosoma asio

Photographs to Aid Identification

SPECIES FOUND IN THE UNITED STATES, CANADA, AND MEXICO

Texas Horned Lizard
Phrynosoma cornutum

Flat-tail Horned Lizard
Phrynosoma mcallii

Coast Horned Lizard
Phrynosoma coronatum

Roundtail Horned Lizard
Phrynosoma modestum

Pygmy Horned Lizard
Phrynosoma douglasii

Desert Horned Lizard
Phrynosoma platyrhinos

Short-horned Lizard
Phrynosoma hernandesi

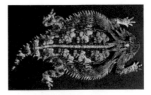

Regal Horned Lizard
Phrynosoma solare

SPECIES FOUND ONLY IN MEXICO

Giant Horned Lizard
Phrynosoma asio

Mexican-plateau Horned Lizard
Phrynosoma orbiculare

Short-tail Horned Lizard
Phrynosoma braconnieri

Bull Horned Lizard
Phrynosoma taurus

Rock Horned Lizard
Phrynosoma ditmarsi

Map of Distribution

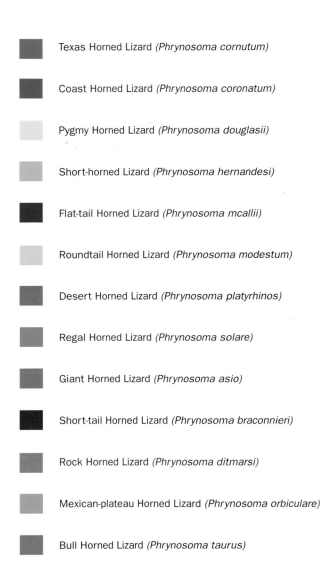

Texas Horned Lizard *(Phrynosoma cornutum)*

Coast Horned Lizard *(Phrynosoma coronatum)*

Pygmy Horned Lizard *(Phrynosoma douglasii)*

Short-horned Lizard *(Phrynosoma hernandesi)*

Flat-tail Horned Lizard *(Phrynosoma mcallii)*

Roundtail Horned Lizard *(Phrynosoma modestum)*

Desert Horned Lizard *(Phrynosoma platyrhinos)*

Regal Horned Lizard *(Phrynosoma solare)*

Giant Horned Lizard *(Phrynosoma asio)*

Short-tail Horned Lizard *(Phrynosoma braconnieri)*

Rock Horned Lizard *(Phrynosoma ditmarsi)*

Mexican-plateau Horned Lizard *(Phrynosoma orbiculare)*

Bull Horned Lizard *(Phrynosoma taurus)*

EIGHT SPECIES FOUND IN
THE UNITED STATES,
CANADA, AND MEXICO

TEXAS HORNED LIZARD   *Phrynosoma cornutum*
Pl. 7

IDENTIFYING CHARACTERISTICS: White midback stripe; two rows of lateral fringe scales; dark lines going up face and over top of head; sharp spines over eyes; two occipital horns pointing upward; ventral scales weakly keeled.

The scientific name of the Texas Horned Lizard is *Phrynosoma cornutum* (from the Latin *cornutus,* meaning horned). It is a large, robust species. The adult generally measures 2.7 to 4.5 inches in snout-to-vent length. Its horns are prominent; the median pair is the longest and these horns are nearly conical in cross-sectional form. The back is very spiny, and there are two rows of enlarged fringe scales along each side of the body. Prominent, nearly erect, spiny scales are scattered over the upper surface. One of these large

Plate 7. Texas Horned Lizard *(Phrynosoma cornutum).*

scales is centered in each of several dark brownish spots on the back. These spots are located in lines along each side of the back and are bordered on their rear margin by cream-colored or yellow crescents. The intensity of colors varies among individuals and populations. The Texas Horned Lizard is unique in having a conspicuous white line, bordered by black, extending down the middle of the back. This is the only species to have dark brown stripes that radiate downward from the eyes to the upper lip and extend upward from the eyes across

Texas Horned Lizard Range

the top of the head. Because the male of this species lacks obvious enlarged scales behind the vent opening (the scales are only slightly enlarged), other characteristics must be used for sexing.

The Texas Horned Lizard is found further east than any other horned lizard in the United States. It inhabits the southern Great Plains, east of the Rocky Mountains, where it can be found throughout much of Texas, Oklahoma, Kansas, southeastern Colorado, New Mexico, and southeastern Arizona. Southward, it occurs in all the bordering Mexican states and

Plate 8. Texas Horned Lizard *(Phrynosoma cornutum)* habitat in Guadalupe Mountains National Park, Texas.

beyond to northern portions of Durango, Zacatecas, and San Luis Potosí. In recent years, the species has been extirpated from areas of eastern and central Texas and portions of Oklahoma. This species occurs in a variety of arid and semiarid habitats in open country having scant vegetation (pl. 8). In these habitats, grasses are often present along with scattered cacti, yucca, mesquite, acacia, juniper, or other woody shrubs and small trees. There are no subspecies of the Texas Horned Lizard.

COAST HORNED LIZARD   *Phrynosoma coronatum*
Pl. 9

IDENTIFYING CHARACTERISTICS: Strong pair of occipital horns, not in contact at bases; bands of dark color crossing back; two rows of lateral abdominal fringe scales; ventral scales smooth; scales on throat pointed.

The scientific name of the Coast Horned Lizard is *Phrynosoma coronatum* (from the Latin *coronatus,* meaning crowned). It is a large species and has a somewhat slender body form. The adult is 2.8 to 4 inches in snout-to-vent length. The Coast Horned Lizard has numerous elongated and pointed scales or spines on the upper surface and two rows of enlarged fringe scales along the sides. Several undulating, blackish brown blotches or transverse bands extend across the back, each highlighted along the rear

Plate 9. Coast Horned Lizard *(Phrynosoma coronatum).*

edge by light cream or tan. A light stripe down the middle of the back is usually indistinct.

The Coast Horned Lizard lives along the Pacific coast of California, west of the deserts and the Sierra Nevada, and southward across the peninsula of Baja California in Mexico. It is found in a variety of habitats—such as chaparral, oak woodland, and coniferous forests—in valleys, foothills, and semiarid mountains from sea level to 6,000 feet in elevation (pl. 10). Increased human population and activity, including the

Coast Horned Lizard Range

Plate 10. Coast Horned Lizard *(Phrynosoma coronatum)* habitat in Joshua Tree National Park, California.

introduction of aggressive exotic ants that are not eaten by horned lizards, have resulted in loss of the species in portions of southern California. In Baja California, the Coast Horned Lizard is found in a variety of arid sections of the Sonoran Desert characterized by strange and beautiful succulent plants. Six subspecies occur throughout the range, including one sometimes considered a separate species *(P. cerroense)*, which lives on Isla Cedros off the west coast of Baja California.

PYGMY HORNED LIZARD   *Phrynosoma douglasii*
Pl. 11

IDENTIFYING CHARACTERISTICS: Adult size small; horns at back of head very reduced, tiny; broad notch extending forward slightly at center back of head; one row of fringe scales; ventral scales smooth.

The scientific name of the Pygmy Horned Lizard is *Phrynosoma douglasii* (*douglasii* honors the collector of the

Plate 11. Pygmy Horned Lizard *(Phrynosoma douglasii)*.

first described specimens of the species, David Douglas, who collected them along the Columbia River in the early 1800s). It is the only species of horned lizard that does not occur in Mexico.

This is a small species, attaining an adult maximum size of 2.5 inches in snout-to-vent length. Until data from recent mitochondrial DNA studies were available, this short-horned lizard was considered one of several subspecies of a more widely distributed species having the older name *P. douglasi* (also spelled *douglassi* and *douglassii*). But now the Pygmy Horned Lizard is considered a distinct species. Collectively, the other five subspecies (individual adults of which are all much larger in size) are now known as the Short-horned Lizard, or *P. hernandesi*. The Pygmy Horned Lizard has extremely reduced horns comprising only small inconspicuous tubercles separated at the midline by a forward-extending hornless space or notch. Dorsal scales are

Pygmy Horned Lizard Range

irregular in size and distribution, and the largest are keeled and surrounded by a rosette of smaller, keeled scales. There is only one row of fringe scales on each side of the body. Two distinct rows of dark blotches or spots extend down the back, with the back edge of each marked by a white or light-colored line.

The range of the Pygmy Horned Lizard extends from south-central British Columbia, Canada (probably extirpated), southward through eastern Washington and Oregon,

Plate 12. Pygmy Horned Lizard *(Phrynosoma douglasii)* habitat in Shasta National Forest, California.

eastward into Idaho, and into northernmost California. It is found in open habitats of high-elevation mesic, or moist, forests and open plains with sagebrush and other shrubs (pl. 12).

SHORT-HORNED LIZARD *Phrynosoma hernandesi*
Pl. 13

IDENTIFYING CHARACTERISTICS: Head horns much reduced in size; deep, hornless notch extending forward at middle of back of head; one row of fringe scales; ventral scales not keeled.

The scientific name of the Short-horned Lizard is *Phrynosoma hernandesi* (*hernandesi* honoring Francisco Hernández Médico, who in 1651 wrote an early

Plate 13. Short-horned Lizard *(Phrynosoma hernadesi)*.

account of a horned lizard). This species attains an adult size of 3 to 4 inches in snout-to-vent length, depending on age and geographic area. Similar to *P. douglasii,* with which it was formerly considered to be a single species, and to *P. ditmarsi,* it has very short head horns separated at the midline by a wide, indented notch. The body is bordered by a single row of fringe scales on either side. There are notable geographic differences in the colors and patterns of individuals from different populations of this highly variable species, the evolutionary relationships of which are not well understood.

The Short-horned Lizard has a broad distribution in the western United States that extends northward slightly into western Canada (Alberta and Saskatchewan) and southward into portions of Sonora, Chihuahua, and Durango, Mexico. Perhaps as a result of its extensive ecologic and geographic distribution, with interrupting continental barriers and glacial history, the Short-horned Lizard has five subspecies. It

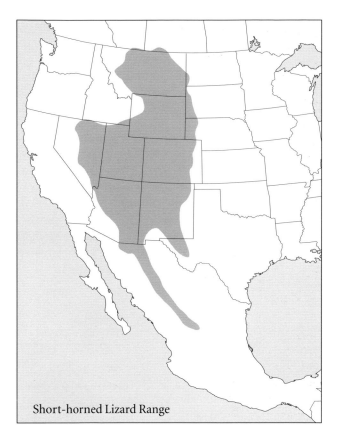

Short-horned Lizard Range

dwells in the short-grass communities of the northern Great Plains, east of the Rockies. In the Great Basin region, it occurs in areas with shadscale, greasewood, and sagebrush. To the south, it populates mountain hillsides and valleys, either in open plant communities such as pinyon-juniper, or in aspen, pine, spruce, or fir forests, where spaces in the canopy allow direct penetration of sunlight (pl. 14). This species occurs as high as 10,400 feet in elevation in the San Francisco Mountains of northern Arizona.

Plate 14. Short-horned Lizard *(Phrynosoma hernandesi)* habitat in Guadalupe Mountains National Park, Texas.

FLAT-TAIL HORNED LIZARD   *Phrynosoma mcallii*
Pl. 15

IDENTIFYING CHARACTERISTICS: Horns long, thin, and sharp; dark line down middle of back; two rows of fringe scales on each side, lower often weak; base of tail dorsoventrally flattened; lip of vent without black spotting; back skin smooth with small spines.

The scientific name of the Flat-tail Horned Lizard is *Phrynosoma mcallii* (*mcallii* honoring the collector of the first described specimen, Col. George A. M'Call of the United States Army, who collected it in California in the early 1850s). It is a medium-sized horned lizard; mature individuals measure 2.5 to 4.3 inches in snout-to-vent length. In this species the two median horns are particularly long and sharp. The chin spines along the lower jaw are also long and narrow. The

Plate 15. Flat-tail Horned Lizard *(Phrynosoma mcallii)*.

Plate 16. Flat-tail Horned Lizard *(Phrynosoma mcallii)* habitat in Algodones Dunes, Imperial County, California.

Flat-tail Horned Lizard Range

Flat-tail Horned Lizard has a double row of fringe scales along each side of the body; scales in the lower row are smaller. This is the only species with a dark vertebral line down the middle of its back. On either side of the sand-colored back are two series of brown spots. The long tail is broad and noticeably flattened. Under the tail, the cream-colored or white front lip of the vent lacks black spots.

The Flat-tail Horned Lizard is found only in the region of the lower Colorado River, in southeastern California and southwestern Arizona and in the adjoining Mexican states of Baja California (northeast) and Sonora (northwest). Over the ages, the Colorado, a major western river, has transported, sorted, and deposited erosional materials throughout its lower basin. Winds have moved and further sorted the fine particles. As a result of these geologic and climatic forces, sand dunes are scattered throughout the area today. These sandy-soil habitats of sparse vegetation are some of the most severe desert habitats occupied by any species of horned lizard (pl. 16). Recent land development for housing, agriculture, and recreation has resulted in a threat to this species within the restricted limits of its distribution. There are no subspecies of the Flat-tail Horned Lizard.

ROUNDTAIL HORNED LIZARD *Phrynosoma modestum*
Pl. 17

IDENTIFYING CHARACTERISTICS: Adult size small; no lateral fringe scales at edge of abdomen, between limbs; four separated, equally spaced horns of uniform size across back of head; black spotting at edge of vent; slender tail round in cross section; back skin smooth.

The scientific name of the Roundtail Horned Lizard, or Bleached Horned Lizard, is *Phrynosoma modestum* (from the Latin *modestum*, meaning calm, unassuming, modest). This is one of the smallest species of horned lizard; adults are 1.6 to 2.7 inches in snout-to-vent length. The four spikelike horns, of nearly uniform size, are distinctive. The bases of these horns are well separated from each other and in a straight line across the back of the head. The Roundtail Horned Lizard is unique in lacking fringe scales along the sides of the abdomen. Its upper surface is not spiny, and it lacks a vertebral stripe; darker spotting patterns are variable. Like several other horned lizards, its coloration differs greatly among populations—it matches local

Roundtail Horned Lizard Range

stones and soils. The Roundtail Horned Lizard has dark blotches above the shoulders, behind the head. In cross section, the tail is round and broadens abruptly at the base. The front tip of the vent is black spotted. The tail is darkly banded throughout its length, as are the legs, both patterns visually isolating them from the "stone" body.

The Roundtail Horned Lizard is native to western Texas, southern New Mexico, and southeastern Arizona and extends into north-central Mexico, mainly in the states of Chihuahua,

Plate 17. Roundtail Horned Lizard *(Phrynosoma modestum).*

Coahuila, Zacatecas, and San Luis Potosí, where it is sometimes known as *tapayatxin*. It occurs in rocky open areas, often in foothills of arid and semiarid basins characterized by desert shrub communities of the Chihuahuan Desert (pl. 18). There are no subspecies of the Roundtail Horned Lizard.

Plate 18. Roundtail Horned Lizard *(Phrynosoma modestum)* habitat in Big Bend National Park, Texas.

DESERT HORNED LIZARD *Phrynosoma platyrhinos*
Pl. 19

IDENTIFYING CHARACTERISTICS: Horns well developed, sharp; single row of fringe scales along sides of body; black spotting at vent opening; color pattern banded; back skin smooth with small spines.

The scientific name of the Desert Horned Lizard is *Phrynosoma platyrhinos* (from the Greek *platy,* meaning flat, broad, or wide, and *rhinos,* meaning nose). This species is medium sized; the adult has a snout-to-vent length of 2.4 to 3.7 inches. The horns are of moderate size and sharp, and the central pair is the longest. Each side of the body carries a single row of fringe scales. The back is marked with wavy blotches or bands that vary in intensity among individuals and populations. Its lighter background color matches the local soils.

Plate 19. Desert Horned Lizard *(Phrynosoma platyrhinos).*

Desert Horned Lizard Range

The Desert Horned Lizard inhabits much of the Great Basin, Mojave, and Sonoran Deserts (including northeastern Baja California and northwestern Sonora, Mexico). Throughout its widespread range, it is associated with woody shrubs, cacti, and yuccas in diverse arid and semiarid habitats on flats, washes, and valleys (pl. 20). There are three described subspecies, but recent mitochondrial DNA studies suggest that the patterns of genetic divergence within the species may not correspond to subspecies designations.

Plate 20. Desert Horned Lizard *(Phrynosoma platyrhinos)* habitat in Death Valley National Monument, California.

REGAL HORNED LIZARD　　　　　*Phrynosoma solare*

Pl. 21

IDENTIFYING CHARACTERISTICS: Flattened crown of horns arranged in a graded series around the back of the head, with bases in contact; four occipital horns (unique to this species); single row of lateral fringe scales; no black spotting at vent; dorsal scales strongly keeled and spiny.

The scientific name of the Regal Horned Lizard is *Phrynosoma solare* (from the Latin *solaris,* meaning of or belonging to the sun). The adult of this large species measures 2.8 to 4.5 inches in snout-to-vent length. Its horns are very distinctive; this is the only species with four central horns, the bases of which are all in contact. These horns grade uniformly into adjacent temporal horns, forming a complete crown and giving the species a regal appearance. The horns are rather flattened and are broader than they are thick. A single row of large fringe scales extends along each side of the body. On the upper surface, numerous large, erect

Plate 21. Regal Horned Lizard *(Phrynosoma solare)*.

scales give this horned lizard a particularly spiny appearance. Body markings are often combinations of browns, grays, and black, with a lighter-colored, large, oval area on the center of the back. It may have a distinct or faint middorsal vertebral line.

The Regal Horned Lizard is found in southern Arizona, in the Sonoran Desert, and eastward barely into the southwestern tip of New Mexico. In Mexico it occurs throughout most of Sonora (including Isla Tiburón) and in northern Sinaloa,

Regal Horned Lizard Range

and a few populations extend eastward into Chihuahua. It is found in arid and semiarid flats and valleys and slopes of mountain foothills. It occurs in desert shrub habitats, some with large, columnar cacti and small trees, and occasionally extends into open oak and juniper woodlands (pl. 22). There are no subspecies of the Regal Horned Lizard.

Plate 22. Regal Horned Lizard *(Phrynosoma solare)* habitat in Saguaro National Monument, Arizona.

FIVE SPECIES FOUND
ONLY IN MEXICO

GIANT HORNED LIZARD

Phrynosoma asio

Pl. 23

IDENTIFYING CHARACTERISTICS: Large in size; with two rows of back spines, one down each side; middle pair of horns (occipitals) usually erect; usually only two temporal horns on each side of head; strong spines above eyes; ventral scales keeled; two rows of lateral fringe scales.

The scientific name of the Giant Horned Lizard is *Phrynosoma asio* (the Latin *asio* meaning horned owl). This may be the largest species of horned lizard in total size and weight, reaching 4.9 inches in snout-to-vent length, with an elongated tail, deep and forward-protruding head, and extremely spiny appearance. In overall body form, this species is the most elongate, thus having a body form slightly more typical of a lizard. Ventral scales are keeled, and there are two rows of lateral fringe

Plate 23. Giant Horned Lizard *(Phrynosoma asio).*

Giant Horned Lizard Range

scales along the sides of the body; the upper row is very well developed. The spines on the head directly above the eyes are long. The central pair of horns at the back of the head protrudes vertically upward, and two horns on each side of the back of the head extend more laterally. On the back, two rows of well-developed spines run down each side from the head to the tail, leaving a relatively spineless central-back area.

The Giant Horned Lizard is found only in semiarid basins of southwestern Mexico on the Pacific side, extending south-east from Colima and southern Jalisco through Michoacán, Guerrero, Oaxaca, and Chiapas. It occurs on rocky hillsides covered by short, seasonally deciduous tropical forests with leaf litter, columnar cacti, and open, sun-exposed semidesert areas of succulent plants (pl. 24).

Plate 24. Giant Horned Lizard *(Phrynosoma asio)* habitat in Guerrero, Mexico.

SHORT-TAIL HORNED LIZARD
Pl. 25

Phrynosoma braconnieri

IDENTIFYING CHARACTERISTICS: Tail very short; ventral scales keeled; single row of lateral fringe scales; horns of moderate length, occipitals separate from temporals.

The scientific name of the Short-tail Horned Lizard is *Phrynosoma braconnieri* (the Latin *braconnieri* honoring the French naturalist Séraphin Braconnier). As the recently acquired common name suggests, the tail is very short, even for the

Plate 25. Short-tail Horned Lizard *(Phrynosoma braconnieri).*

genus, but two other species also have distinctly shortened tails, the Rock Horned Lizard *(P. ditmarsi)* and the Bull Horned Lizard *(P. taurus).* The Short-tail Horned Lizard *(P. braconnieri)* is moderate in size. Head horns are well devel-

Plate 26. Short-tail Horned Lizard *(Phrynosoma braconnieri)* habitat in Oaxaca, Mexico.

Short-tail Horned Lizard Range

oped, but not large; the central pair of horns is slightly larger and separate from the groups of laterally located temporal horns. Ventral scales are keeled.

The habits of this rare species are not well known. It is found in open arid and semiarid basins (pl. 26), including xeric thorn scrub, and at higher elevations in pine-oak woodlands (at approximately 7,000 feet) in the southern Mexican states of Puebla and Oaxaca.

ROCK HORNED LIZARD *Phrynosoma ditmarsi*
Pl. 27

IDENTIFYING CHARACTERISTICS: Tail very short; head horns tiny, reduced to edge of expanded ridge of skull; narrow and deep notch extending forward at back of head; ventral scales keeled; back edge of jaw very deep, expanded; single row of lateral fringe scales.

The scientific name of the Rock Horned Lizard, or Ditmars's Horned Lizard, is *Phrynosoma ditmarsi* (named for Raymond L. Ditmars, early curator of reptiles at the New York Zoological Park and author of numerous and influential editions of *The Reptiles of North America*). After three specimens of this species were collected in Mexico near the United States border in 1890 and 1897, it was not seen again for 73 years until rediscovered in 1970 in Mexico, just south of the border

Plate 27. Rock Horned Lizard *(Phrynosoma ditmarsi).*

with Arizona. Efforts to relocate this mysteriously missing lizard species read like a detective story, with forensic examination of gut contents (species of ants and mineralogy of pebbles) used to geographically locate the origin of the first three specimens.

This species is medium sized, having a snout-to-vent length of up to 3.5 inches. Like two other Mexican species, the Short-tail Horned Lizard (*P. braconnieri*) and the Bull Horned

Rock Horned Lizard Range

Lizard *(P. taurus)* (neither of which occur in Sonora), the Rock Horned Lizard has a very short tail. Both of the other species have much longer cranial horns. The Rock Horned Lizard may have the shortest horns in the genus. The Rock Horned Lizard has been confused with the Short-horned Lizard *(P. hernandesi)*, but the Rock Horned Lizard has shorter horns, a very deeply extended lower jaw, and keeled scales on the ventral side, including the chin. A pronounced notch extends forward at the middle of the back of the head, and the back edge of the skull is nearly a straight ledge across the head. The spines above the eyes are prominent.

The Rock Horned Lizard is known to occur in only four localities in the state of Sonora, Mexico. Habitats are varied and include Madrean evergreen woodland, Sinaloan deciduous forest, and thorn scrub (pl. 28).

Plate 28. Rock Horned Lizard *(Phrynosoma ditmarsi)* habitat in Sonora, Mexico.

MEXICAN-PLATEAU HORNED LIZARD
PI. 29

Phrynosoma orbiculare

IDENTIFYING CHARACTERISTICS: One row of lateral fringe scales; horns not greatly enlarged nor reduced; ventral scales smooth, lacking keels; tail not shortened.

The scientific name of the Mexican-plateau Horned Lizard is *Phrynosoma orbiculare* (from Latin *orbis* meaning circle or circular, here referring to the body form). This species is spiny appearing, medium sized, and adults are 2.5 to 2.7 inches in snout-to-vent length. It has one row of lateral fringe scales, and

Plate 29. Mexican-plateau Horned Lizard *(Phrynosoma orbiculare).*

the ventral scales lack keels. The tail is well developed. The horn patterns consist of lateral groups of temporal horns of about the same length as the two centrally located occipital horns, each of which is neither greatly elongate nor widely separated from the other. Often a tiny spine can be found between the bases of these two horns.

This species is found at high elevations from approximately 4,500 to 11,000 feet and occurs throughout the plateaus and mountain ranges of central Mexico and northward

Plate 30. Mexican-plateau Horned Lizard *(Phrynosoma orbiculare)* habitat in Querétero, Mexico.

along the highland corridors of the Sierra Madre Occidental and Oriental. It has six described subspecies, which are found from eastern Sonora and western Chihuahua southward through western Durango and from Zacatecas and Nuevo León southward to Colima, northern Michoacán, and northern Guerrero, and eastward to Puebla. Its occurrence is largely confined to semiarid shrubland, montane woodland, and forest habitats (pl. 30).

Mexican-plateau
Horned Lizard Range

BULL HORNED LIZARD   *Phrynosoma taurus*
Pl. 31

IDENTIFYING CHARACTERISTICS: Tail very short; horns at sides of head appear fused on elongate temporal region of skull, bull-like; ventral scales keeled; single row of lateral fringe scales; occipital horns greatly reduced; spines above eyes moderately strong.

The scientific name of the Bull Horned Lizard is *Phrynosoma taurus* (*taur* from the Latin meaning bull, ox, or steer). One of the three very short tailed Mexican species, this lizard is distinctive for its laterally extended, elongate, and nearly fused temporal horns behind the eyes and much reduced occipital horns near the midline, which give it a bull-like appearance. Spines above the eyes are prominent. A medium-sized species, the snout-to-vent length is approximately 2.3 to 3.2 inches. Ventral scales are keeled, and it has a single row of lateral fringe scales.

Geographically, the Bull Horned Lizard is restricted to apparently isolated areas in three southern states of Mexico:

Plate 31. Bull Horned Lizard *(Phrynosoma taurus)*.

Bull Horned Lizard Range

Guerrero, Oaxaca, and Puebla. It occurs in varied dry habitats, including seasonally deciduous tropical scrub and semi-arid areas of abundant succulent vegetation with columnar cacti (pl. 32). Although the Bull Horned Lizard generally occurs at lower elevations (4,900 feet) than does the Short-tail Horned Lizard *(P. braconnieri)* (which also occurs in Oaxaca and Puebla), the two species do sometimes occur together.

The fusion of several lateral temporal horns and the reduction of the size of the occipital horns give this species an

Plate 32. Bull Horned Lizard *(Phrynosoma taurus)* habitat in Puebla, Mexico.

outstanding cranial armature. Its appearance is unique. It would be interesting to know which predator actions caused the evolution of such horns, or if it was the mating needs of males holding females's horns in their jaws during copulation.

CONVERGENCE

Sometimes in the course of evolution, unrelated animals independently acquire convergent adaptations that result in remarkably similar structures or forms. The horned lizard, of the New World family Iguanidae, and the Australian Thorny Devil *(Moloch horridus),* of the Old World lizard family Agamidae, are a striking example of convergent evolution. Of similar size, horned lizards and the Thorny Devil are all stocky and spiny (pls. 33, 34). The horned lizards of North America resemble one another because of their close genetic heritage. Horned lizards resemble the Thorny Devil, in spite

Plate 33. Texas Horned Lizard *(Phrynosoma cornutum)* in North America.

Plate 34. Thorny devil *(Moloch horridus)* in Australia.

of their different genetic lineages, because both have independently evolved analogous adaptations for similar lifestyles in analogous environments on separate continents.

Both horned lizards and the Thorny Devil are food specialists, eating large numbers of ants (pls. 35, 36). The Thorny Devil eats almost nothing else, and the ants they eat are tiny. Both lizards extend their tongue and use its sticky end to pick

Plate 35. Regal Horned Lizard (Phrynosoma solare) eating ants in North America.

Plate 36. Thorny Devil (Moloch horridus) eating ants in Australia.

up their prey, in toad fashion. When rain falls, which is infrequently, the Thorny Devil and at least several species of horned lizards use their broad backs to collect water (an act called "rain harvesting") (pls. 37, 38) and channel it to their mouths, which open and close rhythmically, via capillary forces in tiny furrows between body scales. Interestingly, the Thorny Devil even rubs its belly in moist sand so its scales can pull water from between the sand grains into its rain-collection system.

Both types of lizard live in open arid country, and both employ similar strategies to avoid being seen and eaten by

Plate 37. Texas Horned Lizard *(Phrynosoma cornutum)* "rain-harvest" drinking in North America.

Plate 38. Thorny Devil *(Moloch horridus)* "rain-harvest" drinking in Australia.

predators. As with horned lizards, the color pattern of the Thorny Devil blends with its sandy background and incorporates a broken, disrupted pattern, which is insurance against easy detection by visually hunting predators. A common need for camouflage in an arid environment of sparse vegetation has resulted in convergence in appearance. In both types of lizards, the effectiveness of this cryptic coloration is enhanced by their behavior of staying immobile in the presence of danger. Neither is a rapid runner (although the metabolic physiology of the Thorny Devil is slower than that of horned lizards). When wary, the Thorny Devil has a strange gait, not well understood and not seen in horned lizards, that consists of a series of forward and backward rocking motions.

As suggested by their common names, the most striking resemblance between horned lizards and the Thorny Devil is their spiny appearance. Both have strong, sharp horns and spines near the head to deter enemies from swallowing them

Plate 39. Regal Horned Lizard's *(Phrynosoma solare)* spiny head defense in North America.

Plate 40. Thorny Devil's *(Moloch horridus)* spiny neck defense in Australia.

whole. But closer examination of these structures reveals differences that support the contention that horned lizards and the Thorny Devil evolved from separate origins. The Thorny Devil lacks horns around the rear edge of its head; instead, it has a strong spine protruding above each eye and a large, thorny hump projecting above its neck, which horned lizards do not have. If molested, a horned lizard tilts its chin down to raise its horns, and a Thorny Devil tucks its head down between its front legs, leaving its hump, with two lateral spines, as a bite-repulsive pseudohead (pls. 39, 40). The large back scales on the Thorny Devil are spinier and of somewhat different form than those on most horned lizards, but both lizards defensively erect these scales by inflating the lungs. The armor of both kinds of lizards has a similar purpose. Thus, there are both differences and striking similarities between the single Australian species of Thorny Devil and the 13 species in the North American genus of horned lizards.

CYCLES OF ACTIVITIES

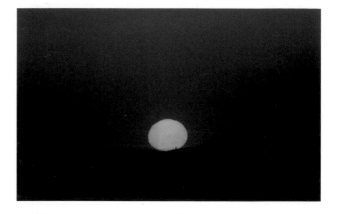

In a lifetime, not every individual completes a full cycle of growth, survival, and reproduction. The reproductive successes and failures of genetically distinct individuals over generations continually adapt a species to changing environments. Thus, each species of horned lizard has been genetically molded by its environment, and each individual inherits a genetic framework for a complex way of life.

Days and Seasons

The rotation of the earth and its revolutions around the sun influence the activities of all horned lizards. Sun rays warm their otherwise cold and immobile bodies, permitting activity.

Horned lizards, along with other reptiles, fishes, and amphibians, are sometimes called "cold-blooded" vertebrates. In contrast, birds and mammals are known as "warm-blooded" vertebrates. But, although it may seem incongruous, a cold-blooded horned lizard sitting in the late-morning sun can have a body and blood temperature equal to, or even higher than, the body temperature of a warm-blooded kangaroo rat resting in its underground burrow. Obviously, the cold-blooded versus warm-blooded terminology can be misleading.

A better distinction between the two groups is that the body temperature of lizards tends to conform to the temperature of their physical environment, whereas the body temperature of birds and mammals remains nearly constant over a wide range of environmental conditions. Basic to this distinction is a difference in the source of energy used to heat the body. Cold-blooded animals utilize direct solar heating, which causes their body temperature to change over a broad range, depending on thermal situations encountered in their environment. Because their heat source is external, they are appropriately called "ectotherms" (*ecto* meaning external; *therm* meaning heat). In contrast, warm-blooded creatures maintain a continually high body temperature that is independent of their physical environment. They accomplish this through the internal metabolic combustion of fuel: energy-rich organic molecules ingested as food. Because their heat source is internal, these animals are referred to as "endotherms" (*endo* meaning within).

Basic physiological differences between ectotherms and endotherms greatly affect the daily activity patterns, food requirements, and annual activity cycles of the two groups. In nature, thermal aspects of the environment are complex and constantly fluctuating—night versus day, dawn versus noon, patches of shade versus areas of open sun, subsurface soil versus surface soil. Horned lizards, like other ectotherms, exploit these differences in order to regulate their body temperature. By orienting their bodies in relation to the sun's rays, moving to warmer or cooler microenvironments, and by varying the amount of time spent in these locations, lizards can behaviorally thermoregulate their body temperatures within narrow limits.

Great energy savings result from the ectothermic method of temperature regulation. Sunning lizards burn five to 10 times fewer calories than resting birds or mammals that are maintaining the same body temperature. The food-fueled, metabolic mode of high-temperature maintenance found in endotherms is costly. For example, the food needed by a small insect-eating bird during one day is sufficient to maintain a lizard of similar weight for 35 days.

By comparing various aspects of ectothermy and endothermy, we see that each method has advantages and disadvantages. Ectothermic lizards have low food requirements, and of the food they eat, a greater proportion is used for growth and reproduction. Endothermic birds and mammals have an internal heating system and are therefore much less restricted in their daily and seasonal activities by time of day, weather, climate, or shade (e.g., habitats with a closed canopy of trees). As a result, endotherms are more successful in overcoming low-temperature limitations to nocturnal activity or to wide geographic distribution in northern regions or at higher elevations.

Also, as a result of their physiology, endotherms can sustain high levels of activity over long periods of time without tiring, an ability that has allowed them to evolve a broad array of sometimes lengthy activity patterns — complex behaviors associated with predation, escape, social interaction, and reproduction. In contrast, ectotherms tire rapidly. Over a short distance, whiptail lizards can run with bursts of speed up to a respectable 5 miles per hour, but they become exhausted in less than 5 minutes when walking continuously at only 0.3 miles per hour.

We should not view the "lower" ectothermic vertebrates as simply unperfected predecessors of the "higher," more successful endothermic vertebrates. In reality, ectotherms are highly successful and precisely adapted exploiters of an alternative, low-energy-demanding lifestyle.

A horned lizard may initiate its daily pattern of thermoregulatory behavior by digging itself out of the loose soil

Plate 41. Morning emergence into the sun for warming (Short-horned Lizard *[Phrynosoma hernandesi]*).

into which it burrowed the previous afternoon or evening (pl. 41). It may emerge shortly before sunrise, probably depending on an internal time sense, or biological clock, to alert it to the dawning day. In leaving its retreat, it may suddenly scamper to the surface and seek an appropriate place to bask in the sun. Alternatively, it may push only its head through the surface, leaving its body and limbs buried. It waits until its body has warmed before exposing itself completely.

A horned lizard with only its head exposed has begun a complex heating process involving the circulation of blood (pl. 42). Its head is warmed directly by the sun. Meanwhile, its body temperature also increases because the surrounding soil is slowly also being warmed by the sun. But as the animal's temperature rises, its head is always 5 to 9 degrees F warmer than its body.

This temperature differential between the lizard's body and its head is partly due to the transfer of heat among large blood vessels whose walls are in contact along their length but that have blood flowing in opposite directions. Warm blood leaving the head in major veins loses some of its heat to cooler

Plate 42. Early morning sunning with only head emerged (Regal Horned Lizard [*Phrynosoma solare*]).

blood passing it in adjacent parallel arteries and entering the head. The transfer of heat from blood leaving the head to blood entering the head circulates heat into the head and keeps the rising head temperature above that of the rising body temperature. This warms the brain rapidly and enables the cool but partially exposed lizard to become alert to dangers more quickly.

Often, just prior to full emergence from the soil, the horned lizard's eyes suddenly bulge out, or protrude, because of an increase in blood pressure, a change caused by tiny muscles surrounding the two major veins that leave the head of the horned lizard. When these muscles are constricted, blood flow from the head is slowed or stopped. Nevertheless, the heart continues to pump arterial blood into the head. As pressure increases in the head, the blocked venous blood is shunted to small veins for return to the body. Thus, it no longer passes next to cooler arterial blood entering the head. These circulatory changes cause body and head temperatures to be brought to equilibrium just before the lizard emerges

Plate 43. Ribs spread and back oriented to morning sun (Regal Horned Lizard [*Phrynosoma solare*]).

from the soil. When it fully exposes itself, it cannot afford to have an alert mind but a sluggish body.

Once above ground, the horned lizard usually begins to bask by tilting its body and turning its back toward the sun (pl. 43). It may face away from the rising sun while standing high on extended forelegs, or it may orient a lowered side toward the eastern horizon while elevating its opposite side by extending its legs on that side. Sometimes, the horned lizard stands with its forelegs on elevated objects or sits on steep, east-facing slopes or rocks. All these tactics increase the angle between the rays of light arriving from the recently risen sun and the horned lizard's back. This directed exposure increases the horned lizard's body temperature more rapidly because a surface held perpendicular to incoming rays absorbs more radiant heat.

While basking, horned lizards increase the amount of blood circulating through the small capillary vessels in the skin on their backs. Heat is rapidly taken up from the skin by the blood and dispersed throughout the body. Horned lizards also spread their ribs and pull them forward (pl. 43), which increases solar heating efficiency by increasing the total area exposed to sunshine.

Early in the morning, the horned lizard alternates between basking and other activities. Basking is discontinued when surface and air temperatures have risen and the higher angle of the sun in the sky enables the lizard to maintain warm body temperatures without basking.

Later in the morning, the ground becomes even warmer and the sun's rays are more direct. Now the horned lizard must prevent its body temperature from rising any higher; relatively slight increases could be fatal. At this point, the horned lizard shuttles back and forth between sunny and shady areas. In the shade, it flattens its body against the cooler ground, losing heat to it. When in the open sunshine, the horned lizard faces the sun so that its back is oriented away from it and pulls its ribs backward to minimize the area of skin exposed on its back (pl. 44).

Throughout the morning, horned lizards defecate, search for food, eat, avoid being eaten themselves, and interact with one another. Continually, they take time out from these activ-

Plate 44. Body compressed while facing into midday sun (Regal Horned Lizard [*Phrynosoma solare*]).

ities to adjust body temperature by increasing or decreasing their exposure to solar radiation.

Direct sunlight can be damaging to living cells. Ultraviolet radiation produces some types of skin cancer in humans. Visible and ultraviolet radiation can penetrate the skin and thin body walls of many desert lizards, including horned lizards, and can damage reproductive and other organs by altering chemical structures and biochemical processes or by inducing mutations. But horned lizards, like several other desert lizards, have an internal protective shield that reduces this solar radiation hazard. In these lizards, the peritoneum, a normally transparent tissue that lines the body cavity of vertebrates, is black and therefore impedes passage of visible and ultraviolet radiation, effectively protecting the internal organs.

By late morning, even shuttling behavior is ineffective for maintaining its body temperature safely below lethal limits, and the horned lizard retreats to the shade of a shrub. Or, it may burrow itself into cooler subsurface soil. The Texas Horned Lizard *(Phrynosoma cornutum)* might even climb up into a bush to pass the hottest part of the day, well above the hot surface soil and heated air next to the ground.

Throughout midday, as the earth turns and the sun moves across the sky, patches of shade shift position. The horned lizard that had earlier burrowed in shaded areas may now find its underground retreat exposed to direct sunshine and heating unbearably. The lizard may burst from beneath the hot sand and run for shaded areas where it rapidly propels itself into the cooler subsurface soil. Of course, if it is burrowed at the central base of a well-shaded shrub, it may not have to relocate when the sun's position changes. The lizard may return to this spot for several days. In some species, a lizard may dig tunnels to pass the day underground, and in others, a lizard may, at midday, sit under the edge of a large rock, secure in the shadows.

To dig itself into the soil, the horned lizard arches its back and lowers its head to the substrate as it rapidly vibrates its head sideways (pls. 45, 46). Strong, pointed scales on both sides of its lower jaw serve as a cutting edge. Meanwhile, the animal's forelimbs are held against the sides of its body (pl. 45). When its head and forebody are submerged, the horned lizard lowers the rest of its body to the ground. Now, waves of lateral oscillations begin passing down the rear half of the body until all of it is buried except the vibrating tip of the tail, which soon disappears. Initially, the horned lizard buries itself to a depth of 0.5 to 1 inch and moves deeper later if necessary. In loose soil it can burrow to depths of 2 to 3 inches in less than a minute.

Plate 45. Burrowing: begins with vibrating the head side to side (Desert Horned Lizard [*Phrynosoma platyrhinos*]).

Plate 46. Burrowing: nearly covered (Desert Horned Lizard [*Phrynosoma platyrhinos*]).

Respiration can be difficult for an animal that is buried underground and surrounded by loose sand and dust particles. With each exhalation, sand drops down into the vacated spaces surrounding the animal's body. On the following inhalation, there is no room to easily expand the chest and fill the lungs. Breathing becomes very labored, as it would in a person buried to the neck in sand at the beach.

The horned lizard avoids this problem by limiting respiratory movements of the chest and abdomen to the underside of the body. With each exhalation, a small, cupped cavity is formed underneath the animal (pls. 47, 48). Downward movement of loose sand into this vacant pocket is blocked

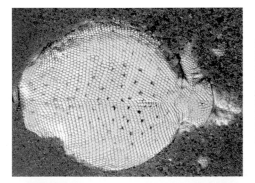

Plate 47. Belly (from below) of lizard burrowed underground, lungs inflated (Regal Horned Lizard [*Phrynosoma solare*]).

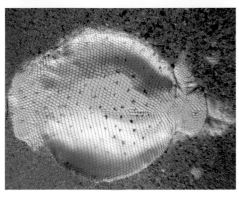

Plate 48. Belly (from below) lof lizard burrowed underground, lungs deflated (Regal Horned Lizard [*Phrynosoma solare*]).

from above by the animal and along the sides by the animal's tucked-under forelimbs. Therefore, the lungs can reinflate after expiration by expanding the chest into the empty cavity below, without having to push back the surrounding soil.

Another potential problem for a buried horned lizard is inhalation of soil particles. However, the design of its nasal openings and passages nicely prevents entry of dirt into the lungs. When underground, the horned lizard elevates a nasal valve from the floor of each nasal tube (pl. 49). This reduces the nasal opening to a tiny crescent-shaped slit, which is still adequate for breathing but prevents entry of most soil particles. Any particles that do pass beyond the nasal valves are trapped in the nasal tubes. Just inside the nasal valves, each of

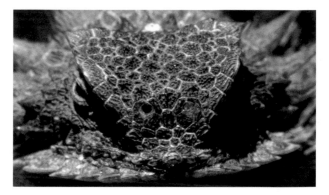

Plate 49. Nasal valves: left open, right closed (Regal Horned Lizard [*Phrynosoma solare*]).

the tubes bends abruptly upward and then sharply downward, in the form of an inverted so-called kitchen-sink trap. In order for an inhaled particle to be sucked further inward, it must be raised straight up against the force of gravity. If it passes this barrier also, it may be blown out of the nasal tube by forced exhalations, or it may become trapped in mucus secreted by glands lining the walls of the nasal tubes.

The skin edges of horned lizard eyelids have a row of tiny receptors that physically detect objects. These are connected to nerves, presumably warning the lizard when to tightly shut its eyelids. Nevertheless, gusty winds sometimes blow grains of sand and soil into the eyes of horned lizards. Also, soil particles occasionally pass between their tightly shut eyelids while they are burrowing themselves underground. These particles in the eyes, unless confined and removed, can damage the cornea.

The horned lizard can wipe intruding particles from the surface of the eyeball with an elastic, transparent sheet of tissue, the nictitating membrane, a structure also present in other lizards. This membrane is usually folded into the front corner of the eye, but it can sweep back over the surface of the eye to pick up foreign debris. The material thus picked up accumulates as a mucus-encapsulated pellet in the rear corner of the eye. This tiny mass of dirt particles grows in size and must be removed. At the base of the nictitating membrane is a blood sinus, a tissue that swells when its numerous capillary vessels are flooded with blood. When muscles around the major blood veins leaving the head are contracted, blood pressure in the eye area of the head increases. This increase fills the blood sinuses around the eye, forcing the enlarged pellet of dirt from its resting site onto the eye or eyelid (pl. 50). From there it either falls off or is deftly removed by a delicate flick of the lizard's hind foot.

Plate 50. Pellet of sand removed from eye by ocular-sinus blood swelling (Regal Horned Lizard [*Phrynosoma solare*]).

By midafternoon, the angle of incidence with which the sun's rays hit the earth decreases, and the surface soil cools and becomes tolerable again for horned lizards. As they adjust to cooling surroundings, their cycle of thermoregulatory behaviors is nearly the reverse of the morning cycle. Emerging lizards tend to remain in the shade at first. As the environment cools further, they shuttle between sun and shade. Later they remain in the sun for longer periods of time and bask in the final rays of sunshine before burrowing again under the loose soil to pass the night. Horned lizards that live on soils unsuitable for burrowing may spend the night on the surface, either in the open or near rocks and small plant cover. The Texas Horned Lizard may climb up into shrubs (pl. 51), possibly to avoid predators such as nocturnal snakes, whereas the Round-tail Horned Lizard *(P. modestum)*, a rock mimic species, prefers to sleep exposed on the ground, near rocks (pl. 52).

Horned lizards modify and adapt their typical daily thermoregulatory pattern to a wide variety of seasonal, geographic, and meteorological conditions. For instance, following emergence from hibernation in early spring, horned lizards often remain active throughout the day. Only later does summer heat force a midday interruption of activity. Temperatures also differ among geographic localities, according to latitude and elevation, and horned lizards respond ap-

Plate 51. Asleep in an acacia shrub a foot above the ground (Texas Horned Lizard *[Phrynosoma cornutum]*).

Plate 52. "Rock mimic" spending the night sleeping next to rocks; radio-transmitter on back (Roundtail Horned Lizard [Phrynosoma modestum]).

propriately to these differences. Further accommodations in their thermoregulatory pattern may be required by weather —clouds, wind, or rain. As ectotherms, their every move is closely attuned to their thermal environment. A favorite site for avoiding midday heat during an intense drought period may not be the same as one used after seasonal rains have begun. With rain, the thermal characteristics of all the micro-habitats the lizard uses change.

The general range in body temperature of active horned lizards is about 84 to 104 degrees F. Below this range, they either find ways to raise their temperature or they become sluggish, seeking shelter and inactivity. During most activities, their body temperature is kept near the upper end of this range. But horned lizards must be careful, for at only slightly higher temperatures, they overheat and die.

Biologists have found that the different species of horned lizards tolerate different body temperatures. The Flat-tail Horned Lizard (P. mcallii) — the species that inhabits the hottest desert — tolerates body temperatures several degrees

Plate 53. Mouth gaping to reduce high temperature through evaporation; radiotransmitter on body (Texas Horned Lizard [*Phrynosoma cornutum*]).

higher before it tries to lower its temperature than do two other desert-inhabiting species, the Desert Horned Lizard (*P. platyrhinos*) and the Roundtail Horned Lizard. To do this, it may seek shade, burrow under sand, discharge cloacal fluids onto its own body (an emergency evaporative-cooling technique), or open its mouth for evaporative cooling (pl. 53). In turn, these two desert-inhabiting species tolerate body temperatures several degrees above those tolerated by two semi-arid or nondesert inhabitants, the Texas Horned Lizard and the Coast Horned Lizard (*P. coronatum*), before they initiate heat-avoidance behavior. Thus it appears that adaptation to differences in high environmental temperature has played a role in molding the behavioral and physiological nature of the various species.

Lizard species thermally adapt to both low and high temperatures. Therefore biologists generally use three ecologically important body temperatures when determining adaptation of a species to the full range of differences in its thermal environment: (1) a preferred body temperature (selected by a lizard in a temperature gradient); (2) a critical-maximum body temperature (at and above which the animal loses muscular coordination and can no longer remove itself from conditions that will soon lead to lethal overheating); and (3) a critical-minimum body temperature (at and below which the animal is too cold and sluggish to walk). By comparing these three temperatures in two species of horned lizards, we can further illustrate the thermal adaptations of species to their environments.

The Texas Horned Lizard inhabits semiarid areas in warmer regions, whereas the Short-horned Lizard *(P. hernandesi)* lives at cool, high mountain elevations and northern latitudes. In experiments, when individuals of the two species are presented with a gradient of temperatures, each species adjusts its body temperature to a different point. The Texas Horned Lizard prefers about 101 degrees F, whereas the Short-horned Lizard prefers 95 degrees F. The critical-maximum temperature of the Texas Horned Lizard is 118 degrees F and that of the Short-horned Lizard is 110 degrees F. The critical-minimum temperature of the Texas Horned Lizard is 49 degrees F and that of the Short-horned Lizard is 37 degrees F. The Texas Horned Lizard can withstand high temperatures that would kill the Short-horned Lizard, whereas the Short-horned Lizard is comfortable and can walk around at lower temperatures than can the Texas Horned Lizard. Its physiological tolerance of low temperatures allows the Short-horned Lizard to live at high elevations and northern latitudes where the Texas Horned Lizard would perish. Each species has adapted to its respective environment, and those adaptations have allowed each species to occupy geographic areas and habitats that are unsuitable for the other.

In addition to movements between sun and shade, horned lizards can alter their color to aid in thermoregulation (pls. 54, 55). Dark lizards sitting in the sun warm faster than do light lizards. Lizard skin has pigment cells, called "melanophores," within which black pigment granules can be moved closer to the skin's surface to darken it or deeper into the skin to lighten it (pls. 56, 57). In dark lizard skin, the black granules in the

Plate 54. Light-colored warm lizard (same individual as in next plate) (Texas Horned Lizard [*Phrynosoma cornutum*]).

Plate 55. Dark-colored cold lizard (same individual as in previous plate) (Texas Horned Lizard [*Phrynosoma cornutum*]).

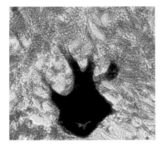

Plate 56. Microscopic view of black pigment in a skin melanophore of a light-colored lizard, showing pigment granules concentrated around the nucleus of the cell (Roundtail Horned Lizard [*Phrynosoma modestum*]).

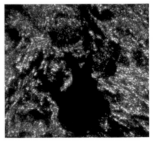

Plate 57. Microscopic view of black pigment in a skin melanophore of a light-colored lizard; light-reflecting cells (iridophores) deeper in the skin dermis are illuminated by polarized light (Roundtail Horned Lizard [*Phrynosoma modestum*]).

melanophores cover other cells (pls. 58, 59), called "iridophores," that reflect light back outward when uncovered by the movement of the granules deeper into the skin. Thus, as a result of upward and downward movement of black granules in the melanophores, the skin color rapidly darkens or lightens. The darkening occurs when a hormone — melanin-stimulating hormone — is released from the brain into the blood and carried to the outside cell surfaces of the melanophores. Here, special receptors recognize the molecules and transmit a biochemical signal to the interior of the cell, where other signals result in a reorganization of the black pigment granules in the cell, moving them toward the skin surface. The skin of the lizard thus darkens, and the lizard absorbs more solar radiation and warms more quickly. By midday, when the lizard avoids heat, it stops producing and dispersing the hormone, and the granules move deeper into the skin, thus exposing the light-reflecting cells and lightening the skin color. Fright may cause the release of other cell-receptor stimulators that also can cause rapid lightening of skin color, as when the horned lizard is captured by a predator.

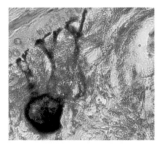

Plate 58. Microscopic view of black pigment in a skin melanophore of a dark-colored lizard, showing pigment granules moved into the branches of the cell toward the outer skin surface, covering the dermis but below the epidermis (Roundtail Horned Lizard [Phrynosoma modestum]).

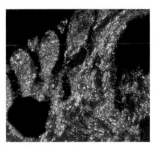

Plate 59. Microscopic view of black pigment in a skin melanophore of a dark-colored lizard; light-reflecting cells (iridophores) deeper in the skin are illuminated by polarized light (Roundtail Horned Lizard [Phrynosoma modestum]).

The skin-color-change responses of three species of horned lizards have been studied: the Texas, Regal *(P. solare)*, and Roundtail Horned Lizards. The first two are large species, and the third is small. The skin of the small Roundtail Horned Lizard has a greater sensitivity to melanin-stimulating hormone than does the skin of the other two species. The Roundtail Horned Lizard also undergoes a greater range of skin-color change (darkening and lightening) than do the other two species. This ability seems to be related to its small size, which, because of its relatively greater surface area to body mass (e.g., a small sphere has a relatively larger surface area in relation to its volume than does a large sphere), enables it to use solar radiation gains more rapidly and effectively than can larger species. Of course this body-size principle applies to the juveniles of all species. Easier daily heating may be a factor in the extended annual activity cycle of juveniles into late fall and early spring.

How does the horned lizard know how much time to spend in the sun each day? The pineal gland, which can be seen through a translucent spot on the top of its head, midway between the eyes (pl. 60), plays some role in regulating the length of time it spends in the sun.

During embryonic growth, the pineal gland develops from the same region of the brain as do the two eyes, and remains

Plate 60. Pineal "eye," white spot with translucent center "lens," on the top of the head (Regal Horned Lizard *[Phrynosoma solare]*).

located midway between them, on the top of the head. Zoologists have long been impressed by the pineal gland's uncanny resemblance to vertebrate eyes. Like true vertebrate eyes, the pineal "eye" has a transparent outer covering, a "lens" for concentrating light rays, and a "retina" on which incoming radiation is focused. In contrast to real eyes, it lacks an iris and fluid-filled chambers. Also, dark pigmentation in the pineal eye is concentrated along the upper surface of the retina, between the source of light and the retinal cells, rather than behind the sensory layer as in functional vertebrate eyes. The pineal gland of the horned lizard also lacks a structure comparable to the optic nerve to connect it to the brain. Whatever the ancestral role of the pineal gland, it is clear that in horned lizards it is not now capable of forming visual images.

Most of our knowledge concerning function of the pineal gland in lizards comes from studies of spiny lizards, close relatives of horned lizards. When the pineal gland was surgically removed from experimental animals, the lizards thermoregulated to the same temperatures as did their unaltered counterparts, but they spent more time sunning. This behavior suggests that one function of the pineal gland in these lizards is to closely monitor their total exposure to sunlight. If they spend too much time in the sun, their energy reserves may be stressed, and their seasonal reproductive cycle may be altered, causing them to be out of phase with other individuals in the population. In truth, this is only a working hypothesis; the full role of the pineal gland, the so-called third eye, in horned lizards remains somewhat of a mystery.

Energy and Growth

Only a minute portion of the sun's massive output of radiant energy shines on the earth. A small percent of this minute portion has fueled life on earth through billions of years of evolutionary change.

Green plants are unique among living organisms in their ability to directly convert solar energy into the chemical energy of molecular bonds. As a result, the energy captured by plants is a force that flows through the bodies of all members of a living community via the energy-transfer system of food chains.

Horned lizards need energy for activity, growth, and reproduction. Like all animals, they obtain it by ingesting and chemically breaking apart energy-rich bonds that link atoms of organic molecules. Their food chains extend back to the sun, which supplies all the energy for every movement of every horned lizard, every day of its life.

The successive links of food chains are held together by energy transfers passing through the line of links. Within any ecological community of organisms, the numerous food chains have beginnings and ends, with similar transitional sections that we recognize as distinct categories, or trophic levels. At the base level of food chains are green plants. Their photosynthetic process captures solar energy and uses it to construct the molecules needed for plant growth. At the second level of food chains are the herbivores: all types of animals (grasshoppers to bison) that eat plants. Herbivores convert the energy stored in plant structures into animal activity and flesh, as well as heat. At subsequent trophic levels of food chains, herbivores are eaten by carnivores, and they in turn are eaten by top carnivores.

In nature, few food chains are simple. Although we can usually assign an organism to a particular trophic level, animals vary their diets and on occasion eat species from various trophic levels. For example, coyotes eat berries, rabbits, and snakes. The diets of some animals change with age and with seasonal availability of food.

The Texas Horned Lizard, like all species of horned lizards, has a decided dietary preference for ants. In particular, this species eats large numbers of harvester ants, which are herbivores that harvest, store, and consume energy-rich seeds of green plants.

All horned lizards eat diets high in ants (50 percent or more), a dietary specialization that is reflected in their simple dentition. Regal and Desert horned lizards eat a diet of nearly 90 percent ants. This dietary preference is easily verified by examination of the contents of one of their large and distinctive-looking fecal pellets, or scats (pl. 61). These pellets almost always contain numerous undigested exoskeletons (outer coverings) of ants. The presence of these pellets on the ground often signals the presence of horned lizards in an area before any have been seen.

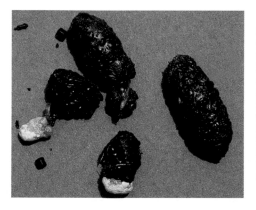

Plate 61. Fecal pellet consisting of harvester ant exoskeleton parts in scat (dark) and white masses of uric acid (Texas Horned Lizard *[Phrynosoma cornutum]*).

When an ant passes near a horned lizard, the lizard may bob its head up and down to stereoscopically determine the distance between it and the ant, take one or two steps forward, tilt its head slightly, and flick out its tongue (pls. 62, 63), upon which it retrieves the ant (pl. 64). It swallows the ant immediately, without chewing.

The Coast Horned Lizard and the Short-horned Lizard have the most varied diets of horned lizard species studied. In addition to ants, they eat beetles, flies, spiders, grasshoppers, moth larvae, and even honeybees. One horned lizard was found feeding at the entrance of a beehive, seemingly unaffected by the numerous stingers implanted in the tissues just inside the margins of its lips.

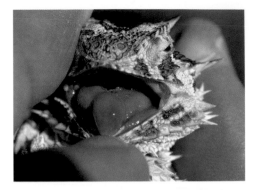

Plate 62. Tongue of an ant-eating lizard (Texas Horned Lizard *[Phrynosoma cornutum]*).

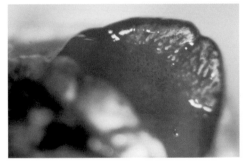

Plate 63. Vascularized tongue tip used to pick up ant prey (Texas Horned Lizard *[Phrynosoma cornutum]*).

Plate 64. Flicking up a seed-harvester ant with the end of the tongue (Regal Horned Lizard *[Phrynosoma solare]*).

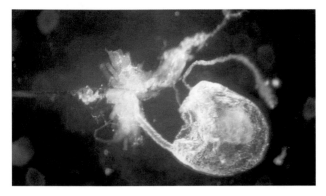

Plate 65. Harvester ant *(Pogonomyrmex rugosus)* stinger (brown) and thin white venom gland and bulbous white storage sac (to right).

Apparently, horned lizards are immune to the highly-venomous stings of harvester ants, common prey. Studies have shown that, in contrast to the blood plasma of a spiny lizard, the blood plasma of a Texas Horned Lizard has the ability to detoxify the venom of harvester ants. This gives them some protection from the ant's venom if stung (pl. 65).

Nevertheless, during swallowing, each live ant, never having touched the teeth of the horned lizard, is coated with a layer of mucus that is secreted by glands in the lizard's pharynx (pl. 66; see mucus on ant legs). This disables the ants, pre-

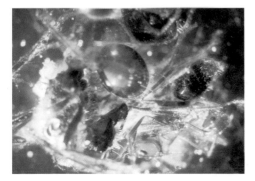

Plate 66. Mucus-covered ants, heads with dark eyes, in the stomach (Texas Horned Lizard *[Phrynosoma cornutum]*).

sumably preventing internal stinging on their way to the stomach where they die in the lizard's digestive fluids. It takes about two days before an individual ant passes through the complete digestive tract and becomes part of a fecal pellet.

In the Chihuahuan Desert of southern New Mexico, there is a valley where two species of horned lizards live together. They both eat ants, but each species prefers different types. In this valley, the Texas Horned Lizard preys almost exclusively on three species of large harvester ants. One lizard may eat 70 to 100 ants in a day. They find these ants in open areas where the ants are searching for seeds, at entrances to ant nests, or along ant paths where columns of ants travel back and forth to gather seeds. The Roundtail Horned Lizard that lives in the same habitat eats honey-pot ants. To catch these ants, the Roundtail Horned Lizard waits in the shade of a small tree or bush for the ants to descend the trunk from the foliage above. The ants are returning from collecting honeydew from sap-sucking aphids and scale insects or collecting sweet plant secretions. In southeastern Arizona, where the same two species of horned lizards occur together but the ant species available are somewhat different, the two lizard species show more overlap in diet, both eating a small seed-harvester ant.

In the Mojave Desert of southern Nevada, the Desert Horned Lizard also eats harvester ants. Here, one species of harvester ant forages in large numbers along straight paths extending 40 or more yards from the nest. At this location, a second species of harvester ant forages individually, never congregating in columns. The Desert Horned Lizard living in the Mojave Desert prefers to eat ants of the species that forages individually. Apparently, its choice is based on previous bad experiences with column-foraging ants that have vicious bites and potent stings.

When a horned lizard approaches a column of this seed-harvesting species, the ants become excited, and several hundred break from their trail to vigorously mob the lizard. Within minutes ants are climbing all over the lizard, biting

Plate 67. Mite larvae in a "mite pocket," a fold in a lizard's skin in front of its foreleg (Regal Horned Lizard [*Phrynosoma solare*]).

into its skin between its belly scales, around its eyes, and between its toes, whereupon the beleaguered lizard flees. In contrast, if a horned lizard is confronted by only one or two potentially biting ants, it simply freezes. Then, if an ant walks over its head, the lizard closes its eyes and often lifts a hind foot to flick off the intruder. Or, it might simply flick out its tongue and eat the ant.

Within any one area, not only do different horned lizard species sometimes find different foods, but the two sexes, being different sizes (the female is generally larger than the male), might exploit slightly different food resources. This tendency is probably highly dependent on local abundance of various prey species. And young horned lizards of the same species may feed on smaller ants and insects, reducing competition with adults.

Some of the energy horned lizards obtain by eating ants can be diverted to the needs of parasitic pests. Tiny ectoparasites live on the skin of some horned lizards, exploiting their host by extracting nutrients and energy. Some of these parasites are the larvae of mites, relatives of ticks and spiders. Many are strikingly orange or red in color (pl. 67). As a rule they congregate in skin folds called "mite pockets" that are located on the sides of the neck of the lizard, just above the front leg. As these larval forms mature, they drop off and become free-living adults.

Nematodes (roundworms) may live in the stomach of the horned lizard, where they extract nutrients and energy from the lizard's meals (pl. 68). As many as 75 percent of horned

Plate 68. Parasitic nematodes and eaten ants in a lizard's stomach (Regal Horned Lizard [*Phrynosoma solare*]).

lizards in some populations have parasitic nematodes. Several hundred may be attached to the stomach wall; over a thousand have been found in one stomach. These roundworms belong to a large group of usually white, worm-shaped animals that lack segments or appendages. Most species of nematodes are free-living in soil or water, but some are endoparasites living within various plants and animals.

Among a typical species of digestive-tract nematode, fertile females constantly discharge eggs into the gut of their host. The eggs, carried out in the feces of the host, may then be eaten by a secondary host and hatch in its digestive tract. Or, in other species, the eggs may hatch before they are eaten, in which case the larvae may burrow into the skin of a host on contact.

How do stomach nematodes travel from one horned lizard to another? The elusive answer was found during a study of the life cycle of a nematode species that infects the Texas Horned Lizard. Somewhat strangely, nematode eggs were not

Plate 69. Cysts enclosing numerous eggs in the two oviducts of a female nematode (*Skrjabinoptera phrynosoma*) (Regal Horned Lizard *[Phrynosoma solare]*).

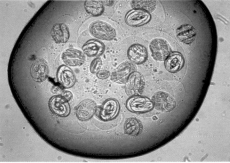

Plate 70. Larval nematodes *(Skrjabinoptera phrynosoma)* in unhatched eggs within a single cyst (Regal Horned Lizard *[Phrynosoma solare]*).

found in the feces of infected Texas Horned Lizards. Also unusually, very few female nematodes living in the digestive tract of these lizards contained fully developed eggs.

Unlike in other nematodes, the eggs in the two oviducts of the female of this particular digestive-tract species develop within resistant capsules (pl. 69), each containing approximately 25 eggs (pl. 70). When a female nematode matures, she does not begin a long period of discharging her eggs into the gut of her host. Instead, she relinquishes her hold on the stomach wall and passes down the intestinal tract and out the lizard's vent (pl. 71). Exposed to the dry air, she desiccates and dies, but her encapsulated eggs, still within the walls of her two oviducts, survive the drying. Each egg, containing a tiny

Plate 71. Female nematode on the surface, discharged with a fecal pellet (Round-tail Horned Lizard [Phrynosoma modestum]).

larva, hatched within the capsule. Now the stage is set for the entrance of the secondary host, through which this nematode must pass before infecting another luckless horned lizard.

Although harvester ants feed primarily on seeds, they are always on the lookout for protein-rich foods to nourish the rapidly growing ant larvae in their nests. A dried nematode seems like the perfect food to feed to these future seed harvesters. So a foraging harvester ant will quickly discard a grass seed in order to pick up the dried carcass of a female nematode, "nematode jerky," to carry back to its nest (pl. 72).

Plate 72. Harvester ant carrying a desiccated female nematode to its nest (Texas Horned Lizard [Phrynosoma cornutum]).

Back in the subterranean ant nursery, the capsule-filled body of the nematode is broken up and fed to larval ants. Inside the moist gut of a larval ant, the capsules open, releasing larval nematodes. Here they grow, passing through several larval stages, while extracting nutrients from the tissues of the ant larva. Later the ant larva pupates and metamorphoses into a young worker ant. Still it carries larval nematodes in its body. When the ant matures, it begins to leave the nest tunnels of the colony to harvest seed. Now, if good fortune strikes the

Plate 73. In the stomach of a horned lizard, a white larval nematode emerges from an eaten ant to infect the lizard (Regal Horned Lizard [Phrynosoma solare]).

larval nematodes and further bad fortune befalls the ant, the ant is eaten by a horned lizard.

Once the ant is swallowed, the larval nematodes it contains emerge as small adults (pl. 73). These attach themselves to the stomach wall of the horned lizard, where they grow and mate. Within 65 days of entering a horned lizard, a female nematode reaches maturity and is ready to cast her body into the cycle again. By neatly fitting its life cycle into the dietary preference of the Texas Horned Lizard, this species of parasitic nematode has ensured its own perpetuation.

Water — obtaining and conserving it — is important for desert lizards, who constantly lose small amounts through their skin and lungs. By burrowing underground, horned lizards help reduce evaporative water loss. They replace water loss with water contained in their foods, and they drink on occasion. They lap up droplets of morning dew from plants, perhaps recognizing them by their metallic shine, and during a rain shower they take water from rock surfaces.

In addition, species such as the Texas, Roundtail, and Desert Horned Lizards move out into rainstorms, whether day or night, and use their backs as rain-harvesting surfaces (pl. 37). Standing in the rain, a horned lizard arches and spreads its back, stands high off the ground on spread legs, and holds its head and tail low. Raindrops falling on its scaly skin are pulled down into the tiny network of spaces between the scales by capillary action, the adhesion of water to the walls of the channels. These same capillary forces pull the water throughout the interscalar network, including to the angle of the jaw at the rear of the mouth. The lizard ingests the water as it slowly opens and closes its jaw in a rhythmic fashion. The flow of water to the edge of the lizard's mouth seems to determine the rate of ingestion, and in heavy rains, gravitational sheet flow over the lizard's back may enhance its drinking rate.

Horned lizards conserve a great deal of water by not using it to flush nitrogenous wastes and salts from their bodies. Their kidneys, unlike those of kangaroo rats, cannot produce a liquid urine solution of higher concentration than the blood. Instead, like most reptiles, lizards eliminate nitrogenous wastes in the form of uric acid. Uric acid is chemically similar to urine but is nearly insoluble in water and therefore can be excreted as a semisolid with little water loss. Water carrying uric acid from the kidney is reabsorbed in the cloaca. From here, the uric acid, mixed with some insoluble crystals of urate salts, is voided as a white mass attached to the end of the fecal pellet (pl. 61).

Some lizards also conserve water by eliminating salts

through their nasal glands. These glands extract excess salts from the blood and secrete them as a very concentrated brine into the nasal passages. The brine may be removed by sneezing. In horned lizards, dried, white encrustations around the nostrils of some individuals (pl. 74) demonstrate that nasal salt glands are present, excreting salts.

With adequate food and water, young horned lizards grow

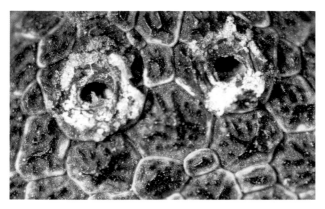

Plate 74. Crystals of salt around both nasal openings (Regal Horned Lizard *[Phrynosoma solare]*).

rapidly. A hatchling Regal Horned Lizard can grow from slightly over 1 inch to a total length of 1.8 inches (snout-to-vent) before going into hibernation. They may remain active until early November, depending on their geographic location. Adults, having accumulated fat reserves for winter dormancy, begin to restrict their movements in late fall and then seek refuge for winter by burrowing underground. In spring, the previous summer's hatchlings may leave hibernation and appear on the surface and begin feeding again before the adults emerge.

These first-year lizards continue to grow very rapidly, utilizing their food energy intake for growth and deferring reproduction until the following year. But some species such as the Roundtail Horned Lizard may reach maturity in one year, especially if climatic conditions result in abundant food. First-year Regal Horned Lizards can grow to a total length of 2.7 inches (snout-to-vent) by early June and to a total of over 3 inches before entering hibernation in fall. The following year they emerge as adults. Thereafter, growth is slower because much of their energy is used in reproduction. But over the years, they can attain a snout-to-vent length of well over 4 inches. The Desert Horned Lizard is known to have lived at least 8 years in the wild. Life expectancy estimates of other species are largely guesswork but may be 2 to 3 years in the Roundtail Horned Lizard, and 5 to 8 years in larger species. Captive individuals of several species have lived over 10 years.

Almost immediately after hatching, young horned lizards shed their outer layer of skin. Thereafter, they molt a few times every year, more frequently as juveniles than as adults. Horned lizards shed their external covering for much the same reason we shed ours: to renew it as it becomes worn and to allow for growth. Unlike us, they shed it from their entire scale-covered body at certain times, although not quite in one piece like a snake. Large irregular flakes of the discarded layer of skin break off over a day or two (pl. 75) during routine activities that flex the skin, exposing the brilliant colors of the new layer of skin that has formed underneath.

The process of removing the old outer layer of skin is problematic around the lizard's inflexible skull. Many iguanid lizards, including horned lizards, increase the blood pressure in their heads, causing slight swelling. The dried outer layer of skin cracks and then flakes off. Head swelling is induced by contraction of muscles surrounding veins leaving the head, reducing the outflow of blood. Arterial blood continues to be pumped into the head by the heart, resulting in the slight swelling. Again, as in cleaning its eyes, the horned lizard's ability to regulate blood flow from its head is useful.

Plate 75. Shedding of the old outer layer of skin in large pieces (Regal Horned Lizard [*Phrynosoma solare*]).

Before entering hibernation, the horned lizard reduces its food intake. Food contained in the stomach while the lizard is underground all winter might decompose. An active horned lizard needs the warmth of the sun's rays and high temperatures to properly digest insects. During hibernation, metabolic processes are reduced; oxygen consumption is low and the rate of respiration decreases. With greatly reduced energy needs, the horned lizard survives the winter months entirely on its reserves of fat and water.

The Texas Horned Lizard in the Chihuahuan Desert of southeastern Arizona remains underground for 6 months of the year. The depth of its hibernacula is only an inch or so under the surface, and the burrow site is located in the open, where it is exposed to daily sunshine and nightly cooling (pl. 76). In the spring, warming soil temperatures may bring this horned lizard out of hibernation. In southern Alberta, Canada, the Short-horned Lizard hibernates for approximately the same time period, October through April, but ex-

Plate 76. Excavated hibernating lizard (with green radiotransmitter attached to back), legs still in burrowing position after 4 months (Texas Horned Lizard [*Phrynosoma cornutum*]).

periences much colder prehibernation and posthibernation days. Nevertheless, it is able to be active in these cold conditions, and its hibernacula are only 3 inches deep, well above the winter frost line. Given these conditions, the Short-horned Lizard may sometimes freeze to death. Nevertheless, this species survives here; its tolerance of this harsh environment is remarkable for a lizard. In contrast, in southern Mexico, the periodic inactivity of horned lizard species is more dependent on seasonal drought and insect inavailability than it is on freezing conditions.

ENEMIES AND DEFENSE

Since horned lizards are potential meals for many animals, they need good defenses to ensure their long survival. Those lizards that survive longest usually leave the largest number of young; they are the parents of tomorrow's horned lizards. Here, in the mathematics of individual differences in lifetime reproductive success, is the means by which natural selection becomes the driving force of evolution.

Lurking Predators

Young horned lizards are the most vulnerable to predators. Being smaller than adults and less well armored, they are more easily captured, killed, and swallowed. Inevitably, many

Plate 77. During the day, Long-nosed Leopard Lizards *(Gambelia wislizenii)* may chase down, kill, and eat small horned lizards (Roundtail Horned Lizard *[Phrynosoma modestum]).*

hatchlings and newborns are eaten before they can enter their first hibernation.

The Long-nosed Leopard Lizard *(Gambelia wislizenii)* frequently kills small, narrow-bodied lizards and swallows them whole, headfirst. It also attacks and eats juvenile horned lizards, as well as the adults of species such as the Roundtail Horned Lizard *(Phrynosoma modestum)* (pl. 77). When a leopard lizard approaches one of these small horned lizards, the horned lizard tries to protect itself by turning its back into a shield, flat and round, and orienting it toward its adversary. The outcome depends on the relative sizes of both lizards, predator and prey. Well-armored adults of larger species such as the Texas Horned Lizard *(P. cornutum)* are not likely to be attacked by predatory leopard lizards, but their hatchling young are.

Many raptorial birds are equipped with strong talons and hooked beaks with which they can kill and dismember horned lizards, allowing them to swallow the lizard's carcass piece by piece. Since large hawks are able to distinguish objects the size of an adult horned lizard from a height of 3,000 feet (the distance of 10 football fields), even these highly camouflaged lizards fall prey to them, as well as to the smaller Prairie Falcon *(Falco mexicanus)*, American Kestrel *(Falco sparverius)*, and Loggerhead Shrike *(Lanius ludovicianus).* Even so, there are some risks to these birds in eating horned lizards. One Red-shouldered Hawk *(Buteo lineatus)* was found dead after having eaten two Texas Horned Lizards; a horn of one lizard had punctured the bird's windpipe.

The Greater Roadrunner *(Geococcyx californianus)* is another threat. A horned lizard can try a valiant defense by flattening its back and aiming it at the bird. But with its extreme speed, the roadrunner can rush a horned lizard and grab its leg or its tail in its beak (pl. 78). Next, the roadrunner hoists the lizard skyward in an oval arch over its head and pulls the lizard back toward the earth in a centrifugal slam against the ground (pl. 79). Repeated poundings soon result in death — and perhaps some broken bones — for the lizard. Then, eating its prey whole with the horns intact, the Roadrunner orients the lizard's body upside down so that when it passes through the bird's expanded throat and esophagus, the horns point away from the bird's lungs, and heart (pl. 80).

Chihauhan ravens *(Corvus cryptoleucus)* bring dead Texas Horned Lizards to the vicinity of their nests to feed to their growing young. But the young birds could die trying to ingest entire horned lizards, so the adults open the lizard's belly with their strong beaks and eviscerate them. Soft parts are fed to the young, and the spiny carcass is discarded.

Snakes, of course, eat their prey whole, killed by either constriction and suffocation or the injection of venom (pl. 81). This is a dangerous undertaking if the meal is a horned lizard. Rattlesnakes are sometimes found dead with the horns of their victim, a half-swallowed horned lizard, protruding through the walls of their throats (pls. 82, 83). Other lizard-eating snakes, such as whipsnakes *(Masticophis* spp.), can suffer the same fate. In contrast, other individuals of the same species of rattlesnakes, such as the Western Diamond-backed Rattlesnake *(Crotalus atrox)* and Sidewinder *(C. cerastes),* have been found to contain half-digested horned lizards in their stomachs — a successful and nutritious meal rather than their last attempt at eating. The difference between life

Plate 78. A Greater Roadrunner *(Geococcyx californianus)* lifts a Texas Horned Lizard *(Phrynosoma cornutum)* off the ground by the lizard's leg.

Plate 79. A Greater Roadrunner *(Geococcyx californianus)* uses a centrifugal slam against the ground to kill a horned lizard (Texas Horned Lizard *[Phrynosoma cornutum]*).

Plate 80. A Greater Roadrunner's *(Geococcyx californianus)* expandable and tough throat allows ingestion and swallowing of a large and spiny meal (Texas Horned Lizard *[Phrynosoma cornutum]*).

Plate 81. A broken-off Western Diamond-backed Rattlesnake *(Crotalus atrox)* fang stuck in back skin shows the cause of death of this lizard (Texas Horned Lizard *[Phrynosoma cornutum]*).

Plate 82. A dead Western Diamond-backed Rattlesnake *(Crotalus atrox)* with the horns of its dead intended victim puncturing its neck (Texas Horned Lizard *[Phrynosoma cornutum]*).

Plate 83. Poor predator judgment by this Western Diamond-backed Rattlesnake *(Crotalus atrox)*, perhaps motivated by acute hunger, led to its death and the death of its prey (Texas Horned Lizard *[Phrynosoma cornutum]*).

and death for both prey and predator seems to be the relative sizes of the individual prey and the individual predator. Proper judgment of the risk involved when capturing and swallowing prey is critical to snake survival.

Levels of Defense

In response to their many enemies with diverse skills, horned lizards have evolved numerous defenses, several of which they may employ in concert. No defense tactic by itself is perfect, and not every defense is appropriate for survival against every predator. Using an appropriate, effective defense during an attack by a particular predator may result in survival, whereas misjudgment may result in death.

For any potential prey animal, we can divide its predator avoidance adaptations into three successive lines of defense. Escape from predators depends first on (1) being unseen, but if that fails, (2) being uncatchable, and if that too fails, (3) being inedible or appearing dangerous. Horned lizards are masters at escaping notice of hungry hunters. If detected, however, they can try running away, but their broad bodies, short hind limbs, and awkward gait are poorly designed for this means of escape from most predators. If captured or cornered, their third line of defense can sometimes be persuasive and life saving. Using defensive behavioral tricks, they can convince some predators that they are better fighters than some want to challenge, that they are too broad, fat, and spiny to be swallowed comfortably, or that they would taste horrible if ingested.

Escaping Notice

We cannot always appreciate the full effectiveness of nature's camouflage when we see an animal in a photograph. Frequently, it is not sitting in its natural background, or its cryptic colors and patterns may be designed for preventing recognition from a greater distance than that used in the photograph. Also, the photographer has already located the animal, centered it in the photograph, and ensured that it is in sharp focus (pls. 84, 85) — all jobs a predator must perform in the field, jobs made more difficult by a horned lizard's cryptic coloration and form.

Predators looking for cryptically colored animals hunt with a search image in mind. Recognition of the shape of an object is very important in vertebrate visual searching. Highway departments have recognized this tendency in people, too. We seek and recognize shapes and have taken this into consideration in the design of traffic warning signs. Disguising

Plate 84. For predators, visually locating a horned lizard requires focusing; here, its eyes are focused on one lizard (Desert Horned Lizard [*Phrynosoma platyrhinos*]).

Plate 85. For predators, visually locating a horned lizard requires focusing; here, its eyes are refocused on another lizard (Desert Horned Lizard [*Phrynosoma platyrhinos*]).

Plate 86. White middorsal stripe on lizard's back blends with dead grass stems (Texas Horned Lizard [*Phrynosoma cornutum*]).

Plate 87. One "rock," lower left, is a rock-mimicking lizard (Round-tail Horned Lizard [*Phrynosoma modestum*]).

Plate 88. Hunching behavior enhances this lizard's rock-like appearance (Round-tail Horned Lizard [*Phrynosoma modestum*]).

shape and color to avoid detection by predators is important to horned lizard survival (pls. 86–90).

The flat bodies of horned lizards are ideal for visually blending into the ground surface upon which they live (pl.

Plate 89. A lizard's cryptic coloration and pattern blend with the pattern of dessicated and cracked desert soil (Desert Horned Lizard [*Phrynosoma platyrhinos*]).

Plate 90. A black mid-dorsal stripe on the lizard's back blends with the plant stem shadows (Flat-tail Horned Lizard [*Phrynosoma mcallii*]).

84). Their behavior when approached confirms the importance of this body form. They immediately flatten their bodies against the soil, hold still, and wait until danger has passed. If the danger is a hawk circling above in the sky, the lizard may tilt its head to keep its enemy in sight, and, when safe, quickly slip under a shrub for increased protection.

Not only does the flat body form of the horned lizard blend into the soil better when flattened even further, but flattening also reduces or eliminates the casting of a body shadow. Even a narrow shadow line along the edge of its body can reveal the horned lizard's form and presence to a predator. Protection against such revealing shadows is provided by the nearly white, lateral-fringe scales found along the lizard's sides. These scales are important to us as an aid in species identification, but for the lizard, they break up and camouflage any dark shadow line that might fall along the body's edge.

One species lacks lateral-fringe scales, perhaps the exception that proves the rule. The Roundtail Horned Lizard lives in rocky and stony areas and looks much like a rock itself. When threatened, it enhances this resemblance by hunching up its back, an act that displays rather than hides its three-dimensionality (pls. 87, 88). And just as rocks have shadows, this horned lizard has darkened areas behind the head, in front of the hind legs, and around the edges of the body, which also uniquely lacks shadow-eliminating fringe scales. Its abdomen mimics the appearance of a rock, and in some populations, the Roundtail Horned Lizard even has darkened pitting shadows that give the illusion of unevenness to its smooth abdomen surface. Individuals differ in color widely within some populations (pl. 91). Varied rock backgrounds promote the evolution of aspect diversity in lizard colors to minimize the facility with which predators can form search images. Here, a predator searching with a one-color search image has poor hunting success. To it, horned lizards appear to be rare meals perhaps not worth valuable hunting time.

In all species, background-color matching aids in escaping detection by enemies. A lizard's color, especially seen from a distance, matches the color of the local soil. These reds and yellows are due to the presence of distinct types of pigment cells in the skin (pl. 92). Background-color matching has been studied in populations of the Flat-tail Horned Lizard (*P. mcallii*) living on slightly different colored sand dunes near the Colorado River. Animals living on the reddish-tinged Algodones Dunes of San Luís, Sonora, Mexico, are redder in overall back coloration than are animals living 70 miles to the north on the whitish dunes of Thousand Palms in Riverside County, California.

Plate 91. Two color forms illustrating some of the variation in color among individuals in one population of a rock-mimicking horned lizard (Roundtail Horned Lizard [*Phrynosoma modestum*]).

Geographical differences in coloration among populations of the Roundtail Horned Lizard can be very striking. The range of this species is transected by the Rio Grande and Pecos Rivers, by numerous streams and valleys, and by the Rocky Mountains. Throughout this area there are several different-colored soils. In New Mexico, east of the Pecos and

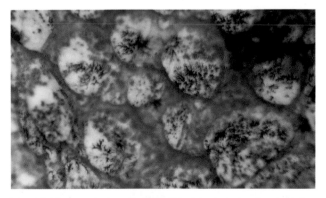

Plate 92. Microscopic view of individual black (melanophore) and red pigment cells spread on top of light-reflecting white cells (iridophores) on magnified back scales (Roundtail Horned Lizard [*Phrynosoma modestum*]).

west of the Staked Plains, alluvial soils are generally red. Between the valley of the Pecos and the southern Rocky Mountains to the west, soils are generally gray. In the Tularosa Basin, soils and rocks are nearly black in the vicinity of lava flows and white near gypsum deposits. In each area, the local Roundtail Horned Lizards match the soil or rock color. Their coloration today has been determined, in a sense, by the geologic and hydrologic forces of the past.

Many horned lizards have dark or black color patterns arranged over their backs (see photographs to aid identification). These patterns help prevent predators from seeing the outline of their animal shape and conformation. On close examination, some of these patterns seem bold and revealing, but when seen from a distance, these same patterns mask the lizard's shape and effectively deceive the predator's eye. Disruptive coloration is as valuable for horned lizard survival as it has been to the survival of many combat soldiers dressed in camouflaged clothing or riding in bizarrely painted tanks.

As part of their disruptive coloration, a few horned lizard species have a striking stripe down the middle of their back (pls. 7, 15, 93, 94). Instead of drawing attention to the animal, this line camouflages it, suggesting a sun-bleached twig or blade of grass. By standing out from the animal's back, the line steals the eye's attention from the less-defined halftones by which the real form of the lizard might be differentiated from the surroundings. The Texas Horned Lizard has a strikingly white midback (pl. 93) line. This species spends much of its time at the base of grasses and small shrubs, where it blends with the stem litter below these plants, and the white line mimics dead plant stems (pl. 86). In contrasting fashion, the Flat-tail Horned Lizard, with its black midback line (pl. 94), lives on sand dunes that are largely windswept clean of plant litter under the sparse plants. Here, this lizard's black line blends with the stem shadows falling on the smooth sand (pl. 90).

The camouflage effect of a good pattern of disruptive coloration is best when certain colors of the pattern closely

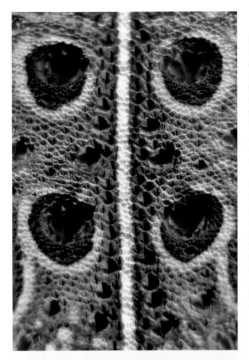

Plate 93. Scales and coloration pattern, illustrating distinctive middorsal white line of the Texas Horned Lizard *(Phrynosoma cornutum).*

Below:
Plate 94. Scales and coloration pattern, illustrating distinctive middorsal black line of the Flat-tail Horned Lizard *(Phrynosoma mcallii).*

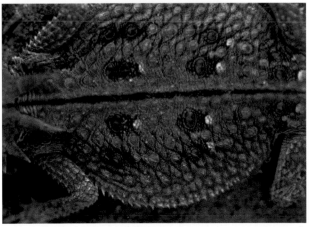

match the background color and other tones strongly differ. This causes some details to blend with the background of the immediate surroundings and other darker and shadowlike details to break the flat surface into three-dimensional relief. The eye sees broken patterns, themselves coherent with the environment, and misses the form of the animal. When the contrast is strong between the tones, the disruptive visual effect is enhanced. Darker patterns on several horned lizard species are directly bordered by brightly colored areas; the transition is sharp, without intermediate gradations or halftones. In this manner, the flat plane of the back is given the appearance of irregular objects at various levels, ridges catching light and valleys left in shadow.

For background-color matching and disruptive coloration to be effective, the horned lizard must behave appropriately. It must remain motionless when in danger of visual detection. Stationary objects, although quite perceptible to our eye and brain, frequently escape our conscious perception. Nothing catches the eye of a searching predator as much as movement against a still background. Horned lizards, relying on their cryptic form and coloration, often remain motionless until nearly stepped on or until picked up. Many horned lizards — unseen by us — must, without moving a muscle, just let us walk by.

When frightened into running, horned lizards run a short distance and then stop abruptly. Some species can scamper quite rapidly with their waddling gait. The brief run and sudden stop has reestablished the horned lizard's first line of defense: invisibility. The predator loses the lizard as it abruptly runs a short distance and then freezes into immobility, and it may never relocate the horned lizard again.

Defending Oneself

Once discovered and cornered, with survival at stake, the horned lizard faces a predator and must convince it of its unpalatability, its third and final line of defense. Most horned lizards have an impressive array of defensive behaviors with which to deter pursuers, but some are largely bluffs. Sometimes the horned lizards are successful, but sometimes it is the predators who are successful.

Recorded responses of horned lizards to humans are diverse. In an almost comic antic, one horned lizard raised itself high on all four legs, lowered its head, and in a series of hops, approached its tormentor like a charging bull. On another occasion, a horned lizard charged and bit a person's boot. Normally they do not bite when captured by hand, but some struggle and attempt to thrust their horns into the flesh of their captor. They can also inflate their bodies with air to make themselves too large to be swallowed. Or they hiss, creating an audio impression of a snake, and with open mouths advertise a false ferocity. Sometimes the horned lizard vibrates its tail nervously. The horned lizard may turn to face a predator approaching its tail end. In one instance, an investigator, after gently handling a Regal Horned Lizard *(P. solare)*, thrust his finger toward the horned lizard's tail. In a single motion, the lizard quickly flipped around in midair, landing in the same spot but pointing in the opposite direction.

But strangely, the horned lizard becomes immobile if picked up and stroked between the eyes or held on its back and rubbed on its belly. After being placed right-side up, one Regal Horned Lizard in this seemingly hypnotic state flipped itself onto its back again. The state usually lasts 5 to 10 minutes, although one individual remained on its back for 2.5 hours. It is not obvious what value this "death feigning" or "playing possum" has for the horned lizard, but children certainly enjoy putting them to "sleep."

In southeastern Arizona, several species of predatory, primarily insect-eating grasshopper mice will kill and eat lizards and even other mice when the opportunity presents itself. A Southern Grasshopper Mouse (*Onychomys torridus*) will attack a sleeping Roundtail Horned Lizard (pl. 95), chewing into its skull above its eye socket. When this mouse attempts the same attack on an adult Texas Horned Lizard, a much larger species, its effort frequently fails and the lizard escapes. The Texas Horned Lizard has a strong spine protruding from a stout bone above each eye. This structure and tough skin can save its life. Grasshopper mice do not attack the array of horns around the back of the head even though the spinal cord area at the back of the head is their favorite chewing target when killing other vertebrates. They seem to recognize its

Plate 95. At night, Southern Grasshopper Mice *(Onychomys torridus)* find, kill, and eat small horned lizards (Roundtail Horned Lizard [*Phrynosoma modestum*]).

invulnerability. Perhaps such a bony defense against chewing predators attacking the base of the skull first led to the evolution of head horns, presumably when they were small in length as in the three present-day short-horned lizard species.

The cranial armament of horned lizards is diverse in the various species. But all species seem to attempt to employ it defensively when molested, pulling their horns up and forward if touched on the snout and eyes (pl. 39). Some species, such as the Flat-tail Horned Lizard, have very long and sharp spines. The potential effectiveness of these spines has been shown by measuring the horn lengths on lizards killed by shrikes and comparing them with the horn lengths in the general population of Flat-tail Horned Lizards. The horns are shorter in the killed lizards, suggesting that predatory birds find longer-horned individuals more dangerous and less desirable as prey.

A snake must consider issues of prey desirability when grasping a horned lizard in its mouth. Of course, different snakes have different hunting strategies. The Western Diamond-backed Rattlesnake waits in hiding, quickly strikes and envenomates its prey, and follows the scent trail to the dead animal, which it can leisurely try to engulf in its spreading jaws. If a Texas Horned Lizard is approached by one of these rattlesnakes, it freezes until the snake is near, then, if it is not sure of the snake's intention, it runs as fast and as far away as possible. Rattlesnakes do not chase their prey, so this works for the lizard. But when faced with another snake, the whipsnake, the lizard's reaction is dramatically different. Whipsnakes feed on the fastest lizards in the desert, chasing them down and grasping them in their jaws. The horned lizard almost never runs when approached by a whipsnake, even if attacked. The snake would chase and quickly grab it from behind, grasping with its jaws above and below the lizard's abdomen. Instead, the horned lizard spreads its back and tilts it upward (on the side away from threat), facing its adversary with a spiny shield. This erect and movable surface is too broad and round for the snake to grasp between its jaws. If the snake attempts to go around the lizard, the lizard shifts position, keeping the shield in the snake's face. When this defense works successfully, the snake tires of trying to get the lizard between its jaws, and not finding a meal, it departs. Horned lizards must make rapid and accurate judgments of predator types to properly assess the attack capabilities of each, and then respond effectively with an appropriate defense.

Plate 96. Swollen, blood-engorged eyelids indicate preparedness for another defensive blood squirt (Regal Horned Lizard [Phrynosoma solare]).

One remarkable feat of most horned lizard species is their ability, when provoked, to shoot narrow streams of blood from their eyes. They shoot blood so infrequently at humans (4 to 6 percent of captures in Texas and Regal Horned Lizards), and reports of it sound so unbelievable, that many people who are familiar with the lizards are misled into regarding blood squirting as a fable. In Mexico, however, many people regard horned lizards as sacred toads. According to folklore, when they cry, they weep tears of blood (pl. 96).

Before a horned lizard squirts blood, it arches its back defensively and closes its eyes. The eyelids become swollen, engorged with blood. Suddenly, a very fine stream of blood, the thickness of a horsehair, shoots out from the slightly parted edges of the closed eyelids. The stream can come from one or both eyes and can spray a distance of 6 feet and be directed forward or backward. Each discharge lasts a second and may be repeated numerous times. A series of blood discharges may be reenacted if the animal is again irritated (pl. 97). A consider-

Plate 97. A blood-squirting defense involves blood loss, but the lizards, if successful, recover quickly (Texas Horned Lizard *[Phrynosoma cornutum]*).

able amount of blood may be lost by the lizard, but it quickly recovers from the experience without obvious ill effects.

Although horned lizards rarely squirt blood at humans, people who handle large numbers of these lizards have observed such a discharge. Humans sampling the blood sprayed do not taste anything strongly objectionable. Why should a horned lizard spray out its own blood? Clues to the answer are found in the circumstances under which this event occurs and the reactions to the blood by potential predators.

Even when vigorously attacked by various predators such as roadrunners, grasshopper mice, leopard lizards, diamondback rattlesnakes, and whipsnakes, the horned lizard does not squirt blood out of its eyes. But, if a blood-shooting species of horned lizard is molested by any doglike species, it often does. If the dog gets the blood in its mouth, it shakes its head vigorously from side to side and shows obvious distaste. Native predators such as the Coyote *(Canis latrans)* and the Kit Fox *(Vulpes macrotis)* react in the same way (pls. 98, 99). Clearly there is some chemical component of the blood that causes dog-family members gustatory (taste) distress. The blood

Plate 98. A Kit Fox *(Vulpes macrotis)* tries a head bite on a horned lizard and receives a squirt of blood in its mouth (Texas Horned Lizard *[Phrynosoma cornutum]*).

never harms them, but it certainly suggests to them that a different meal would be more palatable. How fortunate for the horned lizard to be able to deliver that message before being severely bitten and tasted. Even juveniles of some species can manage a defensive blood squirt when needed. One group of closely related species, however, the Roundtail, Flat-tail, and Desert *(P. platyrhinos)* Horned Lizards (see the diagram of hypothesized relationships), almost never squirt blood. Why these species have apparently lost this defense is a mystery awaiting resolution.

How does the horned lizard manage to squirt blood from its eye sockets? As was mentioned earlier, lizards have a mechanism by which they can increase the blood pressure in parts of their head. There are two sets of constricting muscles surrounding the major veins leading from the head. Upon constriction, these muscles reduce or stop blood flow back out of the head, but arterial blood continues to flow into the head. Under this increased pressure, the ocular sinuses surround-

Plate 99. The Kit Fox *(Vulpes macrotis)* reacts to the squirt-delivered blood (pl. 98) with distaste, shaking its head and later wiping its muzzle in some grass (Texas Horned Lizard *[Phrynosoma cornutum]*).

ing the eyeball become infused with blood and swell, causing the eyelids to bulge. Uniquely in horned lizards under predator attack, the pressure ruptures capillary vessels in the sinuses, and blood flows around the eye, passing the eye membranes and out the tear duct. Contrary to earlier suggestions, no antipredator substances, from glands around the eye, are added to the blood as it leaves the eye socket. Interestingly, on very rare occasions, other lizards under stress have been seen to ooze blood from their ocular sinuses. Perhaps this is how the first horned lizards, with distasteful blood, became the founders of the present-day blood-squirting lineage.

The purpose of blood squirting is defensive against a certain category of predators. Each predator encounter requires the horned lizard to identify the type of predator, which may be a way for the lizard to assess the predator's capabilities and limitations and to determine which defensive methods might be successful. No one defense protects a horned lizard from all of its enemies, and sometimes none work.

COMING OF GENERATIONS

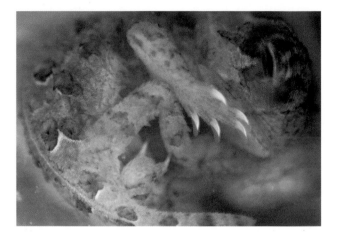

With each new generation, the lines of life are extended another step into the future. In reproduction, the genetic inheritance of the past is united, through natural selection, with change for survival today. Life would never have continued over the more than three billion years it has evolved on earth without the repeated replacement of generations.

Lizard to Lizard

Almost all lizards reproduce sexually. Each is able to distinguish between the sexes and among species, often through social behavior. The function of social behavior is to ensure order in relations among individuals of a species and to allow for the proper selection of mates and timing of mating.

The spatial distribution of individuals in most iguanid lizard populations is a reflection of their social interactions. Typically, males establish territories by defending an area against intrusion from other males of their species by means of threats of aggression or actual physical conflicts that may involve biting. A territorial male frequents favorite perches in his territory, from which he periodically displays rapid up-and-down head-bobbing movements. If other males respond to these signals with head bobbing, they are challenged by the territory holder, who begins increasingly intense push-up displays, which involve repeated flexing and extending of the front legs and prominent exposure of brightly colored body scales (sometimes iridescent blue) on the throat and sides. The intruding male usually leaves, or if a battle ensues, the territorial male is almost always victorious. Body size is also an important factor in establishing territory.

This typical iguanid social arrangement requires that males spend considerable time and energy excluding other males from their territory. In accomplishing this, they risk attracting the attention of predators with their body motions and flashy colors. This risk is balanced by the increased chances territory-holding males have of obtaining a mate or mates for reproduction.

Females consider strong territory holders to possess superior genes that could result in more successful young. Usually, females do not defend territories of their own, and they may frequent the territories of one or more males. When a female is ready to mate, she accepts the sexual advances of a territorial male.

From the little we know about social relations among horned lizards, it appears that they are not as flamboyant in their reproductive behavior as are many of their iguanid relatives. Horned lizards do not display from defended territories, and they lack patches of brightly colored scales. Typically, they roam a familiar area, sometimes termed a "home range," which is undefended. Parts of home ranges may overlap, and the ranges may change over time or seasonally.

Horned lizards are unusual among vertebrates because in nearly all species the females are the larger sex. Females may have evolved a larger body size to increase the number of off-spring they produce, either in number of eggs or live-born young. Males, by becoming reproductively active at a smaller size, can redirect the energy saved from not growing a larger body toward searching for difficult-to-locate females, with the result of fathering more young. Because male horned lizards do not defend territories against invasion by other males, they do not need larger body size for success in ritualized combat. But finding mates is made difficult for horned lizards by their proximity to the soil surface and the consequent intrusion of landscape and vegetation features in their fields of vision. Perhaps they seek open areas such as washes (seasonally dry stream beds) to look for receptive mates.

When two horned lizards first see each other, they enact a rather formal greeting. First one, then the other, bobs its head very slightly several times in rapid succession. The sequence may be repeated several times by each animal. Careful analysis of movies of several other lizard species has shown that the cadence and depth of bobbing differ among species. There is evidence that lizards can distinguish these bobbing patterns and use them, along with other visual cues, to recognize other members of the same species. For horned lizards, we know little about the behavior that follows these initial greetings, but certain observations suggest that they may have complex and intriguing social significance.

For example, one behavioral act appears to involve taste, smell, and sight cues. It occurs after the normal head-bobbing greeting. As one horned lizard approaches another, the stationary lizard reacts by rotating its body so that its tail end faces the approaching lizard. Then it rapidly flashes its tail straight up in the air (pls. 100, 101). This presents the vent and tail base to the oncoming lizard, which extends its tongue and licks the skin of the vent region. The pair quickly vent flash and vent lick. This probably allows the lizard to detect tastes or smells that help it determine the species, sex, state of sexual receptivity, and perhaps even the individual identity of its

Plate 100. Two Flat-tail Horned Lizards *(Phrynosoma mcallii)* learn about each other, one flashing a tail and the other investigating vent chemistry.

Plate 101. A female raises her tail, a male licks her vent for cues to her receptivity for mating (Roundtail Horned Lizard *[Phrynosoma modestum]*).

new acquaintance. Females usually flash their tail to an approaching male, but smaller males may flash to larger males, and smaller females may even flash to larger females.

In a less friendly social interaction, two male Texas Horned Lizards *(Phrynosoma cornutum)* were observed in "combat." First, one lizard grabbed the other with its strong jaws and held firmly. Next, using its horns as weapons, it rapidly lowered and raised its head and in so doing drove its sharp horns into the other male's body. After half an hour of struggling interspersed with rest breaks during which the bitten lizard could not escape, the two were separated. The victim was found to have four bleeding puncture wounds in his throat and another in his chest. The reason for the fight was not clear. Such battles seem to be rare.

For most animals, including horned lizards, spring is the season for mating. But there are some exceptions: some Mexican species mate in fall. Spring mating in horned lizards can occur over a lengthy period, but the length of time of receptivity of an individual female is not well understood. Meetings between an adult male and a receptive female can result in the male rushing the female and climbing onto her back to grasp her head horns in his jaws. During copulation in most species, the male holds the female by her horns toward the back of her head (pl. 102). But in the short-horned species, the males hold onto lateral nuchal folds of skin on the neck just behind the female's ear openings. Both lizards, male and female, twist the tail end of their bodies slightly so that their vents are brought into contact (pl. 103), whereupon the male inserts one of his two hemipenes (pl. 6) into the female's cloaca for transfer of sperm. The mating pair may lie quietly in this position for 15 to 30 minutes before they separate and go their respective ways. Perhaps the most remarkable mating position is seen in the Coast Horned Lizard (*P. coronatum*), in which the pair mates belly to belly. The male turns the female on her back and then holds the folds of skin under her chin firmly in his mouth while inserting his hemipenes.

Eggs

Before entering hibernation, many horned lizards store fats as food reserves in the body cavity. These energy reserves provide for maintenance of slow respiration and blood circula-

Plate 102. A male holds a female by the horns while mating; the female carries a radiotransmitter and antenna on her back (Flat-tail Horned Lizard *[Phrynosoma mcallii]*).

Plate 103. Same pair with tails twisted, which puts their vent openings in contact during mating (Flat-tail Horned Lizard *[Phrynosoma mcallii]*).

tion during hibernation and for initiation of reproduction early the following spring.

Upon emergence from hibernation, many male horned lizards have already utilized some of this energy reserve to enlarge the testes in preparation for sperm production, or, in the case of females, to increase the yolk content of developing ova

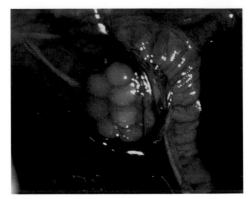

Plate 104. Numerous round ova in the ovary next to the folded oviduct rest on the black internal lining of the abdomen (Regal Horned Lizard [*Phrynosoma solare*]).

(future eggs) in the ovaries (pl. 104). At the proper time, a group of ripe ova are released from the paired ovaries (ovulation) into the open ends of the two tubular oviducts (pl. 104). Within the oviducts, the ova are fertilized by sperm, coated with gelatin-like albumin, and enclosed in leathery eggshells.

The so-called oviductal eggs are carried by the gravid (egg-laden) female (pl. 105) for a period of time during which the multiplication and differentiation of the embryonic cells proceeds in each egg. Higher temperatures speed the develop-

Plate 105. A female's belly bulging with large-shelled eggs in her oviducts, before laying (Regal Horned Lizard [*Phrynosoma solare*]).

mental processes. The embryo is warmed as the sun's rays fall on its mother's back each day. Of course, if the female is killed during this period, the embryos die too.

A female may pass days or weeks before finding conditions appropriate for construction of a suitable nest for her internal living cargo. Site selection is very important for successful incubation of the eggs. Nest soil moisture, drainage, and ventilation, as well as exposure to the sun's warming rays (not being under a shrub), influence the eggs' chances of survival. A female may make several trial digs, perhaps hitting hidden rocks and roots, before being satisfied with a nest site.

Female Texas Horned Lizards and Roundtail Horned Lizards *(P. modestum)* in the Chihuahuan Desert begin this work in the late afternoon or early evening (pl. 106), perhaps so their lengthy and exposed activities do not attract the attention of daytime predators. The female digs a slanted tunnel, semicircular in cross section, into the ground. She digs with her front feet at first, slowly and one at a time. Then as the hole deepens, she kicks accumulated dirt away from and out of the entrance with her hind feet. As the tunnel accommodating her body lengthens, she disappears underground for several minutes at a time.

Plate 106. At night, her tail protruding, a female excavates a nest in the soil to deposit her eggs (Texas Horned Lizard *[Phrynosoma cornutum]*).

A gravid Texas Horned Lizard may expend hours completing a tunnel and hollowing out a small chamber 6 or 8 inches beneath the surface. One horned lizard observed made a two-night project of it. She suspended operations during the intervening day to sit below an adjacent shrub and then resumed digging and laid an estimated 31 eggs. (The nest eggs were eaten by a predator, probably a small snake, before hatching.) During digging, the female is very intent on her activities and will not be deterred from her goal. She can be approached closely and watched. Even after being picked up and then returned to her tunnel opening, she may continue her work.

Plate 107. Excavated underground nest with a group of soil-covered eggs (Regal Horned Lizard *[Phrynosoma solare]*).

Once the nest chamber has been completed, the female begins to lay her eggs in it, one at a time. One female Regal Horned Lizard *(P. solare)* was observed to roll each white egg on the sandy floor as soon as it was laid. This procedure coats the adhesive surface of each egg with a layer of soil identical in color and texture to the surrounding earth and also prevents the eggs from sticking together (pl. 107). Whether this is common in this species or even occurs in other species is not known. The nesting behavior of many species has never been observed. Most Texas Horned Lizards arrange their large clutch of eggs in two or three layers, each level being carefully placed and tightly packed with soil.

As soon as the female has finished laying her eggs, she begins to replace the excavated soil into the nest tunnel. She pushes it in securely with her head as she proceeds up the tunnel. In her efforts to cover the nest hole and disguise its presence, she may scratch the surface for a distance of 3 feet around the nest entrance. When she has buried her eggs, their location is well hidden. Now, she may remain sitting on or near the former tunnel entrance for a day or so. But soon she leaves, never to return or to know if young lizards will hatch from her eggs and emerge to the surface.

The eggs of the larger horned lizard species, the Texas and Regal Horned Lizards, may be 0.7 inches long and 0.5 inches across. Other species lay slightly smaller eggs. Also, females of these larger species lay more eggs per clutch. The Texas Horned Lizard lays 13 to 45 eggs (averaging 31), and the Regal Horned Lizard lays seven to 33 eggs (averaging 21). Other species lay fewer eggs per clutch: the Roundtail Horned Lizard lays six to 18, the Flat-tail Horned Lizard *(P. mcallii)* lays three to 10, the Desert Horned Lizard *(P. platyrhinos)* lays two to 10, and the Coast Horned Lizard lays six to 21. Within some species, the largest egg clutches tend to be laid by the biggest, presumably oldest, females. Interestingly, the Giant Horned Lizard *(P. asio)* of southern Mexico has a reported clutch size of only seven to 21 eggs, which is small in comparison to that of its large northern relatives.

If one female were able to develop and deposit two successive clutches of eggs during a single reproductive season, she would greatly increase the potential number of her offspring. This would require that she repeat the entire egg development cycle, which may take several weeks, after laying her first clutch. In most areas and in many years, this may not be possible because the warm season is short and most horned lizards need to store fat reserves for hibernation and the initiation of reproduction the following spring. But when food is abundant, rainfall is frequent, and weather is mild, the female Roundtail Horned Lizard can produce two clutches in a year.

In Nevada the female Desert Horned Lizard has been observed to lay two clutches of eggs in one year and none in another year, the latter presumably in response to drought and food shortage. Northerly populations of the same species never have more than one clutch in a year. The Coast Horned Lizard, living in the Mediterranean climate of coastal California, is yearly faced with the early onset of summer drought. This has restricted it to a single annual clutch of eggs. The Giant Horned Lizard in Chiapas, Mexico, lays its eggs, which may not hatch for 80 days, in September and October, after the summer rains. From these examples we can appreciate the profound influence of climate, weather, elevation, and geography on the reproductive cycles — which are still not well understood — of each species of horned lizard.

After being transferred from their mother's body to the nest chamber in the upper level of the earth's crust, the embryos in each egg continue to grow. Each day as the sun rises, shines, and sets, letting the soil cool at night, the egg temperature follows a cyclic rhythm. Each day a new generation of horned lizards moves closer to becoming a reality (pl. 108). But if the surrounding soil becomes too dry, the eggs lose water, and the shells of the dehydrating eggs begin to fold. Ants can then bite holes in the folds of the eggshells and destroy the eggs. If still more water is lost from the eggs, the embryos die. Rains percolating down through the soil provide moisture that helps keep the eggs moisture filled, or turgid. But if the nest site was poorly selected, heavy rains may erode the covering soil, exposing the eggs to lethally high temperatures or to predators. Some snake species constantly search for even well-hidden nests of lizard eggs.

Over the weeks, the large yolk in each egg slowly shrinks as the embryo (pl. 109) converts these nutrients into the cells and tissues of a tiny horned lizard. Weather is critical in determining when the eggs will hatch. The sun's warm rays speed

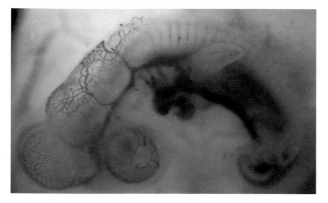

Plate 108. Early embroyonic development of all vertebrates follows a similar pattern: large eye, three brain regions, heart with aortic arches, paddlelike limbs, and segmented torso (Texas Horned Lizard *[Phryno-soma cornutum]*).

Plate 109. Within an egg, an embryo's pigmented eye and pink-ish forelimb, with still-fused digits, are develop-ing (Texas Horned Lizard *[Phrynosoma cornutum]*).

the process. Even so, it may take 5 to 9 weeks before a fully formed young horned lizard is ready to leave its egg-bound world.

All the eggs in a clutch may not hatch. For example, in one nest of 33 Regal Horned Lizard eggs, only 26 hatched. Four eggs disintegrated underground soon after being laid, and three well-formed embryos died before hatching.

Hatching is not easy. The baby lizard has to fight hard to escape from its egg covering (pl. 110). In its efforts, it is aided by a special tool provided for this single moment in its life. That tool is a sharply pointed egg tooth that sticks straight out of its mouth from under the upper lip (pl. 111). With this tooth, the young lizard can puncture and tear the covering that protected it for weeks and now restricts its escape. Once free of the encasing eggshell, the hatchling digs through the soil to the surface. Its digging may be aided by the efforts of a sibling who earlier in the day loosened the soil above the nest on its way to the surface (pl. 112).

By the time hatchlings have reached the surface, they are active miniature versions of the adults. Immediately they begin to head-bob at one another and chase and eat small ants and other insects. When frightened they freeze in position or run for safety.

Live Births

Two closely related species, the Pygmy Horned Lizard (*P. douglasii*) and the Short-horned Lizard (*P. hernandesi*), together have the widest distribution in cooler northern climates of any species of horned lizard. Their ranges include high mountain elevations and northern latitudes. They have adapted their reproductive cycles to very short seasons and low temperatures. Both have foregone laying eggs (oviparity) in underground nests. The cool temperatures of these soils would not permit sufficiently rapid development of the embryos. Instead, both are viviparous, or give birth to live young. Farther south, in Mexico, several species in similar fashion have evolved viviparity. These are the Rock, Mexican-plateau, Short-tail, and Bull Horned Lizards (*P. ditmarsi, P. orbiculare, P. braconnieri, and P. taurus*, respectively). These species all live at higher elevations.

In the two northern species, adults entering hibernation in fall have already increased the size of their testes or

Plate 110. Penny-sized hatchling emerges from egg into its world (Regal Horned Lizard *[Phrynosoma solare]*).

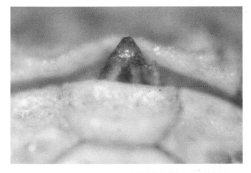

Plate 111. A critical tool for hatching, a temporary "egg tooth," protrudes from between the two lips (Roundtail Horned Lizard *[Phrynosoma modestum]*).

Plate 112. Hatchlings, having left their eggshell underground, emerge above the nest after digging tunnels up through the soil (Texas Horned Lizard *[Phrynosoma cornutum]*).

Plate 113. Head features are seen through a transparent sac surrounding a live-born young as it emerges from the mother's vent (Short-horned Lizard [*Phrynosoma hernandesi*]).

ovaries. This prepares them for mating upon emergence from hibernation in early spring. In spring, the fertilized eggs, without a shell ever being formed around them, are carried inside the female, who serves as a mobile nest. She carries the developing embryos into the sun's rays every day as she basks, changing position as the daily shadow patterns shift, protects them from danger, and maintains the proper moisture level to avoid desiccation. The Pygmy Horned Lizard female stops eating during gestation, perhaps to save the energy that would be used for digestive processes. Also, she is unable to effectively pursue prey. Not eating may also provide abdominal space for developing young, of which there are two to seven. The much larger Short-horned Lizard gives birth to as few as five or as many as 48 tiny horned lizards (the average is about 20), whereas the Short-tail Horned Lizard has four to ten, and one Bull Horned Lizard had thirteen.

The mother may carry the embryos for 2 or 3 months before they are fully developed and ready to be born. Birth appears to be a very casual act involving little or no preparation.

Plate 114. Live-born young emerge from the mother's vent, surrounded by a transparent sac (Short-horned Lizard *[Phrynosoma hernandesi]*).

When the time comes, she simply raises herself high on her hind legs and lets each newborn lizard drop individually from her cloaca (pls. 113, 114). As each one falls, she takes a few steps forward, paying it no further heed. The next birth may come in as little as 1 minute or within 5 or 10 minutes.

Each young lizard is surrounded by a clear liquid and is enclosed within a thin transparent membrane (pls. 113, 114). For the first minute or so after birth, the young lizards remain motionless, but then they begin to wriggle and push against the membrane. After some struggling, interrupted by rest periods, they usually break through the membrane (they too have an egg tooth), allowing themselves and the fluid in which they developed to escape (pl. 115). Then, they draw in their first breath of air. As their lungs fill, their body shape changes from more or less cylindrical to round and flat — like a horned lizard.

Within 30 minutes of birth, these young horned lizards are fully active. Essentially on their own, they must avoid enemies, find food, and build energy reserves before cold weather forces them into hibernation.

Plate 115. Newborn lizard breaks out of its enclosing membranes before taking its first breath (Short-horned Lizard [*Phrynosoma hernandesi*]).

Some of the southern live-bearing species have a slightly different annual reproductive cycle. In the Short-tail and Bull Horned Lizards, mating may take place in fall, after which embryonic development occurs in winter, and birthing takes place in spring. This allows the young many months of growth before cool conditions, with little or no food, return. Other viviparous Mexican species apparently follow the more northern pattern of spring mating, although it is still uncertain whether the Rock Horned Lizard mates in early spring or fall.

OF HUMANS AND LIZARDS

When men and women ventured across the Bering land bridge between Asia and America, perhaps 15,000 to 25,000 years ago, our species entered the New World for the first time. As these Ice Age hunters and their descendants penetrated the American continents, they encountered a truly new world, of which horned lizards were a small but distinctive part.

What were the reactions and thoughts of these earliest immigrants when they encountered lizards with horns and spines? Are they dangerous? Can we eat them? Do their spirits demand respect? Those hunters struggled long ago with their desires, fears, intellect, and emotions, as we do today, to create a view of themselves and of their universe and to develop some understanding of how the two should interact.

Yesterday

Archaeologists believe that by about 2,000 years ago, cultural areas had come to exist within the arid southwestern United States and northern Mexico, including the Anasazi, Mogollon, Hohokam, and Casas Grandes. While passing through stages of development and decline, the peoples of these interacting regions constructed dwellings, some of which are now major archaeological sites. Their artisans employed horned lizard designs in their crafts hundreds of years before Columbus set sail for his discovery of the New World.

The cliff-dwelling Anasazi inhabited and farmed areas of the mesa-strewn Colorado Plateau: Mesa Verde, Chaco Canyon, Canyon de Chelly, Hovenweep, Betatakin, and Keet Seel. The people living in the dwellings at these sites were familiar with horned lizards and painted spiny horned lizards on their pottery (pl. 116) and carved petroglyphs of horned lizards on

Plate 116. A horned lizard design painted on an Anasazi pottery shard (Mesa Verde National Park, Colorado).

Plate 117. Petroglyph of a horned lizard cut into the stone face of a rocky landscape (Zuni, New Mexico).

rocks (pl. 117). About 1,000 years ago, the Mogollon people of the Mimbres River region of southwestern New Mexico developed an exquisite style of black-on-white pottery, upon which they drew many animals, including horned lizards (pl. 118), and scenes from daily life. Until about A.D. 1450, the Hohokam dwelt just south of present-day Phoenix, Arizona, along the Gila River at a site now known as Snaketown. Here, stone carvers sculpted horned lizard effigy bowls and figures (pl. 119). The arrangements of head horns on these sculptures clearly represent two species, the Regal Horned Lizard *(Phrynosoma solare)* (pl. 120) and the Short-horned Lizard *(P. her-*

Plate 118. A horned lizard black-on-white design on a Mimbres-culture bowl from southwestern New Mexico (Laboratory of Anthropology, Santa Fe, New Mexico).

Plate 119. A Hohokam flat stone pallet in the shape of a horned lizard (Arizona State Museum, Tucson, Arizona).

Plate 120. A Hohokam stone lizard sculpture with horn pattern similar to that of a Regal Horned Lizard (*Phrynosoma solare*) (Arizona State Museum, Tucson, Arizona).

nandesi) (pl. 121). Perhaps the Hohokam attributed different powers to the two forms. Some Hohokam craftsmen knew how to etch designs on seashells (pl. 122), which they obtained from the Gulf of California in Mexico. The shell etching was probably done with acid from fermented saguaro fruit, and the unetched portions were covered with protective plant pitch. In the Chihuahuan Desert of northern Mexico, the Casas Grandes people designed bowls with spiny heads and short tails, rimmed by horned lizard fringe scales (pl. 123) or covered with spiny bumps (pl. 124). Further south in Mexico, people were also aware of horned lizards, as evidenced in their designs on pottery vessels and of ceramic seals.

Plate 121. A Hohokam stone lizard sculpture with horn pattern similar to that of a Short-horned Lizard *(Phrynosoma hernadnesi)* (Arizona State Museum, Tucson, Arizona).

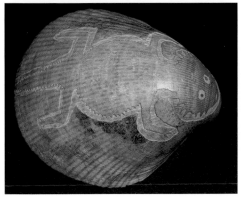

Plate 122. An Arizona Hohokam horned lizard design etched on a seashell from the Gulf of California in Mexico (Arizona State Museum, Tucson, Arizona).

In many cases the descendants of the builders of archaeological sites in North America are among us today. From the lore and myths of Native Americans in the United States and Mexico, perhaps lingering links to the thoughts of their predecessors, we can learn something about these peoples' relationships with the natural world and with horned lizards in particular. In their stories, horned lizards are depicted as ancient, powerful, and respected.

In the 1880s, the ethnographer Frank Hamilton Cushing visited northwestern New Mexico and recorded a Zuni tale of a gifted young hunter who was turned into a mouse by a jeal-

Plate 123. Horned lizard fringe scales and spiny head depicted on a Casas Grandes ceramic pot from Chihuahua, Mexico (Amerind Foundation, Dragoon, Arizona).

Plate 124. Horned lizard head and spiny surface of a Casas Grandes ceramic pot from Chihuahua, Mexico (Amerind Foundation, Dragoon, Arizona).

ous wizard. The wizard tried to take the hunter's wife and home, but a coyote-being came to the rescue. This good spirit dispatched the wizard and took the mouse-charmed young hunter to the underground dwelling of the Great Horned Lizard Medicine Band (pl. 125). In a labyrinth, a number of bag-bellied, grave, old horned lizards, almost as large as men, stirred from their winter sleep to help. One laid the mouse before the altar fire, and another brought a magic crystal and heated it to redness in the embers. Then the master priest and his two warriors spread a sacred blanket over the charmed hunter. Taking their medicine plumes in hand, they dipped

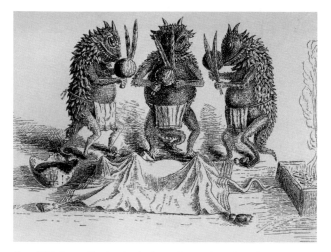

Plate 125. From a Zuni myth, the Great Horned Lizard Medicine Band performs underground magic on a young hunter (from Frank Hamilton Cushing, *Zuni Breadstuff*).

them into a terraced bowl of magic water and sprinkled the shrouded form. Following these preparations, while they chanted and danced, the form under the blanket grew to human size. Then, the master priest snatched the glowing crystal from the fire, and while his assistants raised the blanket, he touched it to the mouse's feet, paws, and head. Momentarily, a transformed young hunter slowly awakened as the horned lizards, chuckling to one another, shrank into the shadows.

Today, Piman peoples living in southern Arizona still recognize the power of horned lizards in their daily lives. They believe that horned lizards can change their fortunes in health and happiness. Among the kinds of sicknesses that Pimans recognize are "wandering sicknesses," those that are contagious afflictions of all races, and "staying sicknesses," those that are not contagious and are restricted to Piman Indians. They believe that staying sicknesses are caused by the "ways" or "strengths" of "dangerous objects" that have been offended. Most of these objects are desert animals such as the horned lizard. At the time of creation, each object was endowed with a dignity, which Pimans must respect. Thus, the Pimans believe that an individual's health is intertwined with the morality of his or her interactions with the natural world.

Horned lizard "strength" is offended if any Piman kills or injures a horned lizard or walks on its tracks, even unknowingly. When this occurs, horned lizard strength enters that person's body. The resulting sickness, however, may not manifest itself until many years afterward, when the person has nearly forgotten the act. It is believed to cause foot sores, and pains or swellings in the feet, legs, hands, and back; it is even believed to cause death.

Plate 126. Tohono O'odahm (Papago) curer's fetish for treating horned lizard sickness (southern Arizona).

When a person is suffering from a staying sickness, a shaman is brought to determine which strength or strengths are causing the sickness and to help the patient recall the incident(s) of transgression. If the shaman determines that a person has horned lizard sickness, ritual curers, persons who know horned lizard songs, are asked to come and sing at the patient's side at night. They may stay for several days. The songs they sing describe horned lizards and how they behave. They make horned lizard strength happy and appeal to its dignity as a being legitimately different from humans. During the singing, a wooden horned lizard fetish (pl. 126), a small carving believed to have magical powers, may be applied to the body or limbs of the afflicted person.

A cure is effected by appealing to horned lizard strength to rid the patient of his or her symptoms. Thus, the way of the horned lizard helps people but only after the horned lizard is shown due respect. This is the essence of the Piman way of living in health and at peace with the natural world.

Today

For about the last 500 years, additional groups of human immigrants have been entering the New World. These people have come mainly from countries across Europe, and also from Africa and Asia. In the United States, Canada, and Mexico, they have come to outnumber the original populations of people. Few of the newcomers had much interest in the beliefs of the native peoples, but as they spread into areas inhabited by horned lizards, they too found these creatures engaging. Representations of horned lizards are increasingly found in the works of modern craftspeople and artists (pls. 127–132).

Plate 127. Horned lizard wood carving (Oaxaca, Mexico).

Plate 128.
Stone carved
fetish (Zuni,
New Mexico).

Below:
Plate 129.
Three-foot-
long wooden
horned lizard
from Mexico.

Some scientists have spent time studying the lives of horned lizards, giving us insights into another life form with which we share the planet. Some people — both children and adults — have tried to keep them as pets, only to discover that they are difficult to keep alive in captivity. Therefore, they are best left where discovered to continue their lives in environments they understand. Also, most states have strict laws against their collection, and the survival of several species is threatened.

Species with small geographical ranges are sometimes at greater threat. The Flat-tail Horned Lizard *(P. mcallii)* living

Plate 130. Horned lizard "priest," ceramic sculpture.

Plate 131. Ceramic pottery design of horned lizard from Mata Ortiz, Chihuahua, Mexico.

in sand dune areas near the lower Colorado River in the United States and Mexico is currently threatened. Its ancestral habitat is being lost to the development of human-managed habitats. In southern Mexico several species have small geographical ranges in regions where human activities are expanding. Little is known of the impact of these human activities on horned lizards, but there is reason for concern because natural habitats are disappearing at alarming rates.

Plate 132. Twin-spouted Navajo wedding vase, covered with ceramic horned lizards.

As backyard residents, horned lizards enjoy a popularity almost equal to that of cardinals, doves, and quail. Unfortunately, admirers sometimes inadvertently contribute to the demise of their backyard horned lizards by ridding their yards

Plate 133. Male Texas Horned Lizard (*Phrynosoma cornutum*) possibly scanning the visual expanse of a road for a potential mate.

Plate 134. Traffic fatality of a Texas Horned Lizard (*Phrynosoma cornutum*) on a highway.

of the biting and stinging ants that horned lizards depend on for food. Also, our ubiquitous blacktop highways and streets are unintentional death traps for horned lizards, either crossing the wide vistas or using them to seek mates (pls. 133, 134). Mechanized and chemically supported agricultural development can bring an end to a place and a community of plants and animals upon which horned lizards have been dependent for everything in their lives. And, our casual worldwide travel and movement of products threatens their lives as well.

Decades ago, Fire Ants *(Solenopsis invicta)* were inadvertently introduced to North America through the port of Mobile, Alabama. These South American ants have been hugely successful colonizers in humid portions of the southern United States. Not only have they disrupted human activities, but they have been a factor in reducing and threatening the existence of native wildlife, including the Texas Horned Lizard *(P. cornutum)* in eastern portions of its range. In southern California, another introduced ant, the Argentine Ant *(Linepithema humile),* has spread widely and become the dominant ant in parts of this landscape. The presence of both aggressive ant species has led to a decline in native ants, the food of horned lizards. The horned lizards in California and Texas refuse to eat either of the introduced species. These ants' recruitment behavior for attacks and chemical defenses eliminate them as possible horned lizard prey. Thus, the Texas Horned Lizard is no longer found in many areas of its former range in eastern and central Texas, and the Coast Horned Lizard *(P. coronatum)* is similarly experiencing range reduction. These changes are the result of inadvertent ecosystem changes brought about by modern human activities. They are only small parts of much wider patterns of human-induced ecological changes being experienced on the planet today.

Before 1945 most people never worried much about the radiation released by atomic decay. But since then, apprehension has grown. In 1964, the U.S. Atomic Energy Commission began supporting an experiment in southern Nevada designed to assess the long-term effects of low-level gamma radiation on natural populations of desert animals and plants. The 22-acre experimental area and its plant and animal inhabitants were continuously irradiated by exposure to radioactive cesium. At that time, radiobiologists predicted that because the exposure rate was low, there would not be any effect on resident animals. But after several years of careful monitoring, it was determined that female Desert Horned Lizards *(P. platyrhinos)* were being sterilized by the cumulative effects of the experimental radiation. The ovaries of females of other lizard species were also being destroyed.

These examples of declines in horned lizard populations should warn us of the unsuspected complexities in interactions between our civilization and other life on our planet. For today, we too, like the Anasazi, Mogollon, Hohokam, Casas Grandes, and other peoples of the past in North America, are struggling to formalize an ethic to guide our relationships to the natural world.

Tomorrow

As the North American wilderness dwindled with the advances of Western civilization, far-sighted people saw that some of it must be saved as a place of meditation and a source of eternal inspiration. Every future generation will face its own struggle of reshaping its ethics to new realities and perceptions. No greater wisdom could be given to future generations than that which awaits contemplation in the integrated beauty of nature.

Horned lizards seem to have a special ability to remind us of our innate connections to the natural world. Our fascination with the lives of each species of horned lizard is but a connecting path to all of life. Can they help us to understand that all our families extend back to the beginnings of life on earth? Do we want to change the ecology of our planet in ways that will lead to the extinction of our biological brethren? The histories of all species on this planet are intertwined and interlinked like the strands and base pairs of a DNA molecule. So are our fates tomorrow.

Time and a place for reflection are indispensable if we are going to see that the mother of all life twinkles in the eyes of horned lizards too. For the story of horned lizards is not an isolated one. Ask yourself: Are they not, as are we, but another color in the rainbow of life shining out of the past through the prism of time and onward into the unknown future?

SELECTED REFERENCES

Bahr, D.M., J. Gregorio, D.L. Lopez, and A. Alvarez. 1974. *Piman shamanism and staying sickness (Ká:cim múmkidag)*. Tucson: University of Arizona Press.

Baur, B.E. 1986. Longevity of horned lizards of the genus *Phrynosoma*. *Bulletin of the Maryland Herpetological Society* 22:149–151.

Baur, B., and R.R. Montanucci. 1998. *Krötenechsen*. Offenbach, Germany: Herpeton, Verlag Elke Köhler.

Bryant, H.C. 1911. The horned lizards of California and Nevada of the genera *Phrynosoma* and *Anota*. *University of California Publications in Zoology* 9:1–70.

Bundy, R.E., and J. Neess. 1958. Color variation in the round-tailed horned lizard, *Phrynosoma modestum*. *Ecology* 39:463–477.

Cushing, F.H. 1874 (1920). *Zuñi breadstuff*. Museum of the American Indian, Indian Notes and Monographs 8. New York: Heye Foundation.

Ditmars, R.L. 1951. *The reptiles of North America*. Garden City, N.Y.: Doubleday and Company, Inc.

Donaldson, W., A.H. Price, and J. Morse. 1994. The current status and future prospects of the Texas horned lizard *(Phrynosoma cornutum)* in Texas. *Texas Journal of Science* 46:97–113.

Funk, R.W. 1981. *Phrynosoma mcallii* (Hallowell), flat-tailed horned lizard. *Catalogue of American amphibians and reptiles 281.1–281.2*. St. Louis, Mo.: Society for the Study of Amphibians and Reptiles.

Gans, C., et al., eds. 1969–98. *Biology of the reptilia*. 19 vols. Vols. 1–13, London: Academic Press. Vols. 14–15, New York: John Wiley & Sons. Vol. 16, Ann Arbor, Mich.: Branta Books. Vols. 17–18, Chicago: University of Chicago Press. Vol. 19, St. Louis, Mo.: Society for the Study of Amphibians and Reptiles.

Heath, J.E. 1964. Head-body temperature differences in horned lizards. *Physiological Zoology* 37:273–279.

Heath, J.E. 1965. Temperature regulation and diurnal activity in horned lizards. *University of California Publications in Zoology* 64:97–136.

Hodges, W.L. 1995. *Phrynosoma ditmarsi* (Stejneger), rock horned lizard. *Catalogue of American amphibians and reptiles 614.1–614.3.* St. Louis, Mo.: Society for the Study of Amphibians and Reptiles.

Howard, C.W. 1974. Comparative reproductive ecology of horned lizards (genus *Phrynosoma*) in southwestern United States and northern Mexico. *Journal of the Arizona Academy of Science* 9:108–116.

Huey, R.B., E.R. Pianka, and T.W. Schoener, eds. 1983. *Lizard ecology: Studies of a model organism.* Cambridge: Harvard University Press.

Jennings, M.R. 1988. *Phrynosoma cerroense* (Stejneger), Cedros Island horned lizard. *Catalogue of American amphibians and reptiles 427.1–427.2.* St. Louis, Mo.: Society for the Study of Amphibians and Reptiles.

Jennings, M.R. 1988. *Phrynosoma coronatum* (Blainville), coast horned lizard. *Catalogue of American amphibians and reptiles 428.1–428.5.* St. Louis, Mo.: Society for the Study of Amphibians and Reptiles.

Jones, K.B. 1995. Phylogeography of the desert horned lizard *(Phrynosoma platyrhinos)* and the short-horned lizard *(Phrynosoma douglassi):* Patterns of divergence and diversity. Ph.D. diss., University of Nevada, Las Vegas.

Lee, S.H. 1955. The mode of egg dispersal in *Physaloptera phrynosoma* Ortlepp (Nematoda: Spiruroidea), a gastric nematode of Texas horned toads, *Phrynosoma cornutum. Journal of Parasitology* 41:70–74.

Lee, S.H. 1957. The life cycle of *Skrjabinoptera phrynosoma* (Ortlepp) Schulz, 1927 (Nematoda: Spiruroidea), a gastric nematode of Texas horned toads, *Phrynosoma cornutum. Journal of Parasitology* 43:66–75.

Lowe, C.H., M.D. Robinson, and V.D. Roth. 1971. A population of *Phrynosoma ditmarsi* from Sonora, Mexico. *Journal of the Arizona Academy of Science* 6:275–277.

Manaster, J. 1997. *Horned lizards.* Austin: University of Texas Press.

Medica, P.A., F.B. Turner, and D.D. Smith. 1973. Effects of radiation on a fenced population of horned lizards *(Phrynosoma platyrhinos)* in southern Nevada. *Journal of Herpetology* 7:79–85.

Middendorf, G.A., III, and W.C. Sherbrooke. 1992. Canid elicitation of blood-squirting in a horned lizard *(Phrynosoma cornutum)*. *Copeia* 1992:519–527.

Middendorf, G.A., III, W.C. Sherbrooke, and E.J. Braun. 2001. Comparison of blood squirted from the circumorbital sinus and systemic blood in a horned lizard, *Phrynosoma cornutum*. *Southwestern Naturalist* 46:384–387.

Milne, L.J., and M.J. Milne. 1950. Notes on the behavior of horned lizards. *American Midland Naturalist* 44:720–741.

Milstead, W.W., ed. 1967. *Lizard ecology: A symposium*. Columbia: University of Missouri Press.

Montanucci, R.R. 1981. Habitat separation between *Phrynosoma douglassi* and *P. orbiculare* (Lacertilia: Iguanidae) in Mexico. *Copeia* 1981:147–153.

Montanucci, R.R. 1987. A phylogenetic study of the horned lizards, genus *Phrynosoma*, based on skeletal and external morphology. *Contributions in Science No. 390*. Los Angeles: Natural History Museum of Los Angeles County, pp. 1–36.

Montanucci, R.R. 1989. Maintenance and propagation of horned lizards *(Phrynosoma)* in captivity. *Bulletin of the Chicago Herpetological Society* 24:229–238.

Montanucci, R.R., and B.E. Baur. 1982. Mating and courtship-related behaviors of the short-horned Lizard, *Phrynosoma douglassi*. *Copeia* 1982:971–974.

Montanucci, R.R., and J.P. O'Brien. 1991. Mating behavior of the coast horned lizard *(Phrynosoma coronatum)*. *Vivarium* 3:27–28.

Munger, J.C. 1984. Home ranges of horned lizards *(Phrynosoma):* Circumscribed and exclusive? *Oecologia* 62:351–360.

Munger, J.C. 1984. Long-term yield from harvester ant colonies: Implications for horned lizard foraging strategy. *Ecology* 65:1077–1086.

Munger, J.C. 1984. Optimal foraging? Patch use by horned lizards (Iguanidae: *Phrynosoma*). *American Naturalist* 123:654–680.

Nabhan, G.P. 2003. Singing the turtles to sea: the Comcáac (Seri) art and science of reptiles. Berkeley: University of California Press.

Norris, K.S., and C.H. Lowe. 1964. An analysis of background color-matching in amphibians and reptiles. *Ecology* 45:565–580.

Parker, W.S. 1974. *Phrynosoma solare* (Gray), regal horned lizard. *Catalogue of American amphibians and reptiles 162.1–162.2*. St. Louis, Mo.: Society for the Study of Amphibians and Reptiles.

Pianka, E.R. 1986. *Ecology and natural history of desert lizards: Analysis of the ecological niche and community structure.* Princeton: Princeton University Press.

Pianka, E.R. 1991. *Phrynosoma platyrhinos* (Girard), desert horned lizard. *Catalogue of American amphibians and reptiles 517.1–517.4.* St. Louis, Mo.: Society for the Study of Amphibians and Reptiles.

Pianka, E.R., and W.S. Parker. 1975. Ecology of horned lizards: A review with special reference to *Phrynosoma platyrhinos. Copeia* 1975:141–162.

Pianka, E.R., and H.D. Pianka. 1970. The ecology of *Moloch horridus* (Lacertilia: Agamidae) in western Australia. *Copeia* 1970:90–103.

Pianka, E.R., and L.J. Vitt. 2003. *Lizards: Windows to the evolution of diversity.* Berkeley: University of California Press.

Pough, F.H. 1969. The morphology of undersand respiration in reptiles. *Herpetologica* 25:216–223.

Pough, F.H., R.M. Andrews, J.E. Cadle, M.L. Crump, A.H. Savitzky, and K.D. Wells. 1998. *Herpetology.* Upper Saddle River, N.J.: Prentice Hall.

Powell, G.L., and A.P. Russell. 1998. The status of short-horned lizards, *Phrynosoma douglasi* and *P. hernandezi,* in Canada. *Canadian Field-Naturalist* 112:1–16.

Presch, W. 1969. Evolutionary osteology and relationships of the horned lizard genus *Phrynosoma* (family Iguanidae). *Copeia* 1969:250–275.

Price, A.H. 1990. *Phrynosoma cornutum* (Harlan), Texas horned lizard. *Catalogue of American amphibians and reptiles 469.1–469.7.* St. Louis, Mo.: Society for the Study of Amphibians and Reptiles.

Prieto, A.A., Jr., and W.G. Whitford. 1971. Physiological responses to temperature in the horned lizards, *Phrynosoma cornutum* and *Phrynosoma douglassii. Copeia* 1971:498–504.

Reeder, T.W., and R.R. Montanucci. 2001. A phylogenetic analysis of the horned lizards (Phrynosomatidae: *Phrynosoma*): evidence from mitochondrial DNA and morphology. *Copeia* 2001:309–323.

Reeve, W.L. 1952. Taxonomy and distribution of the horned lizards genus *Phrynosoma. University of Kansas Science Bulletin* 34:817–960.

Rissing, S.W. 1981. Prey preferences in the desert horned lizard: Influence of prey foraging method and aggressive behavior. *Ecology* 62:1031–1040.

Roth, V.D. 1997. Ditmars' horned lizard *(Phrynosoma ditmarsi)* or the case of the lost lizard. *Sonoran Herpetologist* 10:2–6.

Schmidt, P.J., W.C. Sherbrooke, and J.O. Schmidt. 1989. Detoxification of ant *(Pogonomyrmex)* venom by a blood factor in horned lizards *(Phrynosoma)*. *Copeia* 1989:603–607.

Sherbrooke, W.C. 1981. *Horned lizards: Unique reptiles of western North America.* Globe, Ariz.: Southwest Parks and Monuments Association.

Sherbrooke, W.C. 1987. Defensive head posture in horned lizards *(Phrynosoma;* Sauria; Iguanidae). *Southwestern Naturalist* 32:512–515.

Sherbrooke, W.C. 1988. Integumental biology of horned lizards *(Phrynosoma).* Ph.D. diss., Tucson: University of Arizona.

Sherbrooke, W.C. 1990. Predatory behavior of captive greater roadrunners feeding on horned lizards. *Wilson Bulletin* 102:171–174.

Sherbrooke, W.C. 1990. Rain-harvesting in a lizard, *Phrynosoma cornutum:* Behavior and integumental morphology. *Journal of Herpetology* 24:302–308.

Sherbrooke, W.C. 1991. Behavioral (predator-prey) interactions of captive grasshopper mice *(Onychomys torridus)* and horned lizards *(Phrynosoma cornutum* and *P. modestum).* *American Midland Naturalist* 126:187–195.

Sherbrooke, W.C. 1993. Rain-drinking behaviors of the Australian thorny devil (Sauria: Agamidae). *Journal of Herpetology* 27:270–275.

Sherbrooke, W.C. 1997. Ditmars' horned lizard, or rock horned lizard: An historical update since rediscovery (1970). *Sonoran Herpetologist* 10:6–8.

Sherbrooke, W.C. 1997. Physiological (rapid) change of color in horned lizards *(Phrynosoma)* of arid habitats: Hormonal regulation, effects of temperature, and role in nature. *Amphibia-Reptilia* 18:155–175.

Sherbrooke, W.C. 1999. Thorny devils and horny toads. *Nature Australia Magazine* 26(6):54–63.

Sherbrooke, W.C. 2000. *Sceloporus jarrovii* (Yarrow's spiny lizard): ocular sinus bleeding. *Herpetological Review* 31:243.

Sherbrooke, W.C. 2001. Do vertebral-line patterns in two horned lizards (*Phrynosoma* spp.) mimic plant-stem shadows and stem litter? *Journal of Arid Environments* 50:109–120.

Sherbrooke, W.C. 2002. Seasonally skewed sex-ratios of road-collected Texas horned lizards *(Phrynosoma cornutum). Herpetological Review* 33: 21–24.

Sherbrooke, W.C. 2002. *Phrynosoma cornutum* (Texas horned lizard): Nocturnal nesting, eggs, nest predation, hatching. *Herpetological Review* 33: 206–208.

Sherbrooke, W.C. 2002. *Phrynosoma modestum* (round-tail horned lizard): Death due to prey (beetle) ingestion. *Herpetological Review* 33:312.

Sherbrooke, W.C. 2002. *Phrynosoma modestum* (round-tail horned lizard): Rain-harvest drinking behavior. *Herpetological Review* 33:310–312.

Sherbrooke, W.C., and S.K. Frost. 1989. Integumental chromatophores of a color-change thermoregulating lizard, *Phrynosoma modestum* (Iguanidae: Reptilia). *American Museum Novitates* 2943:1–14.

Sherbrooke, W.C., and M.D. Greenfield. 2002. *Phrynosoma hernandesi* (short-horned Lizard): Defensive hiss. *Herpetological Review* 33: 208–209.

Sherbrooke, W.C., and D. Lazcano Villarreal. 1999. Los camaleones de México. *México Desconocido* 271:50–57.

Sherbrooke, W.C., and G.A. Middendorf, III. 2001. Blood-squirting variability in horned lizards (*Phrynosoma*). *Copeia* 2001:1114–1122.

Sherbrooke, W.C., and R.R. Montanucci. 1988. Stone-mimicry in the round-tailed horned lizard *Phrynosoma modestum* (Sauria: Iguanidae). *Journal of Arid Environments* 14:275–284.

Sherbrooke, W.C., and R.B. Nagle. 1996. A dorsal intraepidermal mechanoreceptor in horned lizards (*Phrynosoma:* Phrynosomatidae: Reptilia). *Journal of Morphology* 228:145–154.

Sherbrooke, W.C., E.R. Brown, and J.L. Brown. 2002. *Phrynosoma hernandesi* (short-horned lizard): Successful open-mouthed threat defense. *Herpetological Review* 33: 208.

Smith, H.M. 1946. *Handbook of lizards: Lizards of the United States and Canada.* Ithaca, N.Y.: Comstock Publishing Co.

Smith, H.M., and E.H. Taylor. 1950. An annotated checklist and key to the reptiles of Mexico exclusive of the snakes. *Smithsonian Institution United States National Museum Bulletin* 199:1–253.

Stebbins, R.C. 1948. Nasal structure in lizards with reference to olfaction and conditioning of the inspired air. *American Journal of Anatomy* 83:183–221.

Stebbins, R.C. 1985. *A field guide to western reptiles and amphibians,* 2d ed. Boston: Houghton Mifflin Company.

Suarez, A.V., J.Q. Richmond, and T.J. Case. 2000. Prey selection in horned lizards following the invasion of Argentine ants in southern California. *Ecological Applications* 10:711–725.

Tollestrup, K. 1981. The social behavior and displays of two species of horned lizards, *Phrynosoma platyrhinos* and *Phrynosoma coronatum. Herpetologica* 37:130–141.

Van Devender, T.R., and C.W. Howard. 1973. Notes on natural nests and hatching success in the regal horned lizard *(Phrynosoma solare)* in southern Arizona. *Herpetologica* 29:238–239.

Vitt, L.J., and E.R. Pianka, eds. 1994. *Lizard ecology: Historical and experimental perspectives.* Princeton: Princeton University Press.

Whitford, W.B., and W.G. Whitford. 1973. Combat in the horned lizard, *Phrynosoma cornutum. Herpetologica* 29:191–192.

Zamudio, K.R. 1996. Ecological, evolutionary, and applied aspects of lizard life histories. Ph.D. dissertation, Seattle: University of Washington.

Zamudio, K.R. 1998. The evolution of female-biased sexual size dimorphism: A population-level comparative study in horned lizards *(Phrynosoma). Evolution* 52:1821–1833.

Zamudio, K.R., and G. Parra-Olea. 2000. Reproductive mode and female reproductive cycles of two endemic Mexican horned lizards *(Phrynosoma taurus* and *Phrynosoma braconnieri). Copeia* 2000:222–229.

Zamudio, K.R., K.B. Jones, and R.H. Ward. 1997. Molecular systematics of short-horned lizards: Biogeography and taxonomy of a widespread species complex. *Systematic Biology* 46:284–305.

Zug, G.R., L.J. Vitt, and J.P. Caldwell. 2001. *Herpetology: An introductory biology of amphibians and reptiles.* 2d ed. New York: Academic Press.

ADDITIONAL CAPTIONS

PAGE I Horned lizard carving on the edge of a Hohokam stone bowl (Arizona State Museum, Tucson, Arizona).

PAGES II–III Visual and chemical signals are used during mate selection by two Roundtail Horned Lizards *(Phrynosoma modestum).*

PAGE VI Summer monsoon rain clouds build over the Chiricahua Mountains in Arizona, habitat of the Short-horned Lizard *(Phrynosoma hernandesi).* Chihuahuan Desert habitat of the Texas Horned Lizard *(Phrynosoma cornutum)* and Roundtail Horned Lizard *(Phrynosoma modestum),* foreground.

PAGES XIV–1 Flattened against the earth's soil and spread by vertebrate ribs, a Desert Horned Lizard *(Phrynosoma platyrhinos)* basks in the sun's rays.

PAGE 4 Aeolean sand-dune habitat of the Flat-tail Horned Lizard *(Phrynosoma mcallii)* in the Algodones Dunes, west of Glamis, California.

PAGE 9 Cleared and stained skeleton of a Regal Horned Lizard *(Phrynosoma solare).*

PAGE 18 Spring poppies cover San Simon Valley foothills habitat of the Texas Horned Lizard *(Phrynosoma cornutum)* and Roundtail Horned Lizard *(Phrynosoma modestum),* foreground. The Chiricahua Mountains in Arizona, habitat of the Short-horned Lizard *(Phrynosoma hernandesi),* are in the distance.

PAGE 20 Three illustrations of fringe scales: top, Roundtail Horned Lizard *(Phrynosoma modestum);* middle, Regal Horned Lizard *(Phrynosoma solare);* bottom, Texas Horned Lizard *(Phrynosoma cornutum).*

PAGES 26–27 Male and female Desert Horned Lizard *(Phrynosoma platyrhinos)* mate to perpetuate their species.

PAGE 28 Open desert shrub habitat of the Texas Horned Lizard *(Phrynosoma cornutum)* and Roundtail Horned Lizard *(Phrynosoma modestum)*, foreground. Higher elevations with woodland and forest habitats in the distant Chiricahua Mountains are home to the Short-horned Lizard *(Phrynosoma hernandesi)*.

PAGE 52 Cactus-rich, xeric habitat of the Bull Horned Lizard *(Phrynosoma taurus)* near Zapotitlán Salinas, Puebla, Mexico.

PAGE 67 Habitat of the Australian Thorny Devil in central Australia at the Kata Tjuta Mountains, its Dreamtime mythological home according to Aboriginal peoples.

PAGES 72–73 Hatching of a clutch of eggs of the Regal Horned Lizard *(Phrynosoma solare)*.

PAGE 74 Sunrise, Sentinel, Arizona.

PAGE 109 Texas Horned Lizard *(Phrynosoma cornutum)* squirting blood from its eye as a defensive behavior. © Raymond A. Mendez.

PAGE 130 Texas Horned Lizard embryo *(Phrynosoma cornutum)* in egg next to remaining yolk.

PAGE 147 Ceramic horned lizard "being," created by artist Susan Nagota Bergquist.

INDEX

Page references in **boldface** refer to the main discussion of the species.

Series Design: Barbara Jellow
Design Enhancements: Beth Hansen
Design Development: Jane Tenenbaum
Cartographer: Bill Nelson
Composition: Impressions Book and Journal Services, Inc.
Text: 9.5/12 Minion
Display: Franklin Gothic Book and Demi
Printer and binder: Everbest Printing Company

ABOUT THE AUTHOR

Wade C. Sherbrooke grew up on Staten Island, before the Verrazano-Narrows Bridge connected it to the rest of New York City (1940s and 1950s). Here, where he played as a child, he found hardwood deciduous woodlands with redbacked salamanders, ponds visited by painted and snapping turtles, blackberry thickets visited by box turtles, and marshes filled with birds and mammals. Today, the area is covered by houses filled with the accoutrements of modern society, as well as city parks. While in grammar school, raising and racing pigeons gave his early interest in animals a focus. By high school he became a summer volunteer under Carl Kauffeld at the Staten Island Zoo (reptile wing), which prompted him to pursue a career in biology. Four undergraduate years as a vertebrate zoology major at Cornell University (B.S.) and a summer fighting fires in Oregon for the U.S. Forest Service prepared him to go west to Tucson, Arizona, for graduate work with reptiles. The rich research environment and talented graduate students at Chuck Lowe's field-oriented ecology program at the University of Arizona molded and shaped his growing passion to engage in the natural world in more intellectual ways and led to herpetology studies in the Grand Canyon (M.S.). Then, breaking for cross-cultural adventures, he spent two years as a Peace Corps volunteer in Amazonian Peru—teaching at Universidad Agraria de la Selva, Tingo Maria, and studying tropical lizard reproduction—before returning to the United States to begin and then drop out of two Ph.D. programs. In 1976, while working at the Office of Arid Lands Studies (University of Arizona) for the San Carlos Apache tribe and studying the Sonoran Desert shrub jojoba, he became interested in horned lizard biology. His studies led to a popular book, *Horned Lizards: Unique Reptiles of Western North America* (1981), and to his return to graduate school at the University of Arizona to study horned lizards. He received his Ph.D. and finally escaped city life by moving to the Chiricahua Mountains of southeastern Arizona (1985) to become the Director of the Southwestern Research Station of the American Museum of Natural History, a position he has grown in and with while studying diverse aspects of horned lizards' lives. He and his wife, Emily, have two sons, Sky and Reed.

CALIFORNIA NATURAL HISTORY GUIDES

"It's always good to read a new California Natural History Guide;
these little books are small enough to fit into a pocket, inexpensive,
and authoritative."   *—Sunset*

"A series of excellent pocket books, carefully researched, clearly
written, and handsomely illustrated."   *—Los Angeles Times*

The California Natural History Guide series is the state's most
authoritative resource for helping outdoor enthusiasts and
professionals appreciate the wonderful natural resources of their
state. If you would like to receive more information about the
series or other books on California natural history, please fill in this
card and return it to the University of California Press or register
online at www.californianaturalhistory.com.

Name _____

Address _____

City/State/Zip _____

Email _____

Which book did this card come from? _____

_____

Where did you buy this book? _____

_____

What is your profession? _____

UNIVERSITY OF CALIFORNIA PRESS
www.ucpress.edu

Return to:
University of California Press
Attn: Natural History Editor
2120 Berkeley Way
Berkeley, California 94720